Doors

THE BEST OF
Fine Homebuilding

DOORS

THE BEST OF
Fine Homebuilding

The Taunton Press

Cover Photo: Charles Miller

Taunton
BOOKS & VIDEOS

for fellow enthusiasts

©1995 by The Taunton Press, Inc.
All rights reserved.

First printing: 1996
Printed in the United States of America

A FINE HOMEBUILDING Book

FINE HOMEBUILDING® is a trademark of The Taunton Press, Inc., registered in the U.S. Patent and Trademark Office.

The Taunton Press, Inc.
63 South Main Street
P.O. Box 5506
Newtown, Connecticut 06470-5506

Library of Congress Cataloging-in-Publication Data

Doors : the best of Fine homebuilding.
    p.   cm.
    Includes index.
    ISBN 1-56158-126-7
    1. Doors — Design and construction — Miscellanea.
    I. Taunton Press.   II. Fine homebuilding.
TT2278.D59   1996
694'.6 — dc20   95-47081
                                                            CIP

Contents

7 Introduction

8 Choosing a Front Door
Steel and fiberglass offer insulation and low maintenance, but engineered lumber has eliminated many of wood's problems

14 Connecticut River Valley Entrance
Reproducing a famous 18th-century doorway

20 A Greek Revival Addition
Joining stock and custom-milled trim to make a formal entryway

24 Formal Entryway
A challenging adaptation of traditional designs

29 French-Door Retrofit
Opening up a wall to bring the outdoors in

34 Trimming the Front Door
Ripped, mitered and reassembled interior moldings add a custom look to a front entry

38 Making an Insulated Door
This handsome entry keeps heat in while it keeps breezes out

40 Batten Doors
Building a solid door from common lumber

42 Replacing an Oak Sill
Doing the job on a formal entry without tearing out jambs and trim

45 A Breath of Fresh Air
Making a Victorian screen door

48 Building Wooden Screen Doors
Durability and aesthetics hinge on sound joinery and steadfast materials

52 Production-Line Jamb Setting and Door Hanging
Time-saving techniques from the tracts

57 Hanging an Exterior Door
From framing the rough opening to mortising for hinges, installing a door requires patience and precision

60 Ordering and Installing Prehung Doors
Precise results require careful planning

64 Installing Prehung Doors
An accurate level and a bucketful of shims will correct just about any out-of-plumb condition

70 Hanging Interior Doors
An organized approach to a demanding job that must be done right to be done at all

77 Casing a Door
Work carefully, and save the wood filler for the nail holes

80 Building Interior Doors
Using a shaper to produce coped-and-sticked frames and raised panels

84 Pueblo Modern
With knife and brush, a furnituremaker applies Southwestern motifs to a houseful of doors

87 Pocket Doors
Should you buy a kit or build your own?

92 Curved Doors
A design with a circular closet in the center of the house calls for some special techniques

95 An Elegant Site-Built Door
Build a custom interior or exterior door with common job-site tools and readily available materials

100 Installing Mortise Locksets
Whether you drill and chisel or rout with an expensive lock mortiser, cutting a big hole for the case is only part of the job

104 Installing Locksets
You can do the job freehand with a chisel and a drill, but jigs and routers are faster

110 Homemade Hardware
Turning scrap metal into hinges, hooks and grilles

113 Door Hardware
Getting a handle on locksets, latches and dead bolts

118 Finishing Touches

126 Index

INTRODUCTION

YEARS AGO, when I still earned my living as a carpenter, my boss sent me out to hang the front door in a new house. We hadn't built the house, so I wondered why we'd been hired to hang the door. I assumed that the construction process hadn't gone well.

When I arrived at the job, the first thing I noticed was that the house was well built. And just inside the entry was a beautiful staircase. I asked the owner, "Who built the stairs?" and he replied that one of the finish carpenters had built them. "Why didn't he hang this door?" I wanted to know. "He said he didn't know how," the owner told me.

As surprising as that was, what happened next upset me even more. The owner came into the foyer with a folding aluminum lawn chair—the kind with woven plastic webbing—sat down in it and watched me for the next four hours while I hung the door.

Hanging a door can be tricky. So can choosing a door, building one from scratch, trimming the opening and installing door hardware. These aren't the sorts of tasks you want to tackle without advice from experts, which happens to be what you're holding in your hands.

This book is a compilation of articles from *Fine Homebuilding* magazine. And as such it represents the collective experience of over 20 builders from all around the country. So whether you're a professional builder, a dedicated amateur or simply a homeowner trying to get a door to swing correctly on its hinges, if you're looking for good information on doors, this book is a likely place to find it.

—Kevin Ireton, editor

Choosing a Front Door

Steel and fiberglass offer insulation and low maintenance, but engineered lumber has eliminated many of wood's problems

by Rich Ziegner

Steel door, integral raised molding. For improved security and appearance, Pease's High Profile Ever-Straight doors have 23-gauge steel skins stamped with molding that projects from the door surface. This prehung door, with brass-cane glazing, costs about $1,130.

Veneer over engineered lumber. Wood-door manufacturers are turning to engineered products, which are more stable than solid wood. This Simpson Sommerset 4002 has a Douglas-fir veneer surface over laminated stiles and rails; it costs about $935.

It's not wood. Therma-Tru's Fiber Classic fiberglass door looks so good that we featured it on the cover of *FHB* #63. The door requires a new UV-inhibiting top coat only every two or three years.

There goes the security alarm, a shrill whistle like a tea kettle come to a boil. The steel door on the west side of the building must have tripped it again. After all, the sun's shining this afternoon, but it's cool inside *Fine Homebuilding*'s editorial offices. Perfect conditions for thermal bow.

What's thermal bow? It's what happens to an exterior door when there's a big difference between indoor and outdoor temperatures. The sun roasts the door's exterior face, but the interior face is cold. Something's got to give; in this case it's the door's shape, which goes from relatively flat to something on the order of, say, a side view of Bill Clinton.

Should the building's planners have chosen a different door? Maybe a wood door or a fiberglass door? Having spent a month researching all three kinds of entry doors, I can tell you, simply, no. There are many good reasons to use a steel door, but there also are compelling reasons to go with wood or fiberglass.

Wood looks good—What makes a wood door so appealing? We've all heard the cliche, "the warmth of wood," and it's true that wood is relatively warm to the touch. But wood's grain patterns also create subtle variations of light and dark colors. Once you've been sensitized to this difference, you can spot a fiberglass or steel pretender because it's flat and monochromatic.

And how about the heft of wood? Wood doors are the heaviest of the three materials, giving wood doors a feeling of quality. A six-panel hemlock door weighs about 10 lb. more than a similarly styled fiberglass door and about 15 lb. more than a six-panel steel door.

Because wood can be worked so easily, wood doors also offer the widest range of styles. The two basic forms are flush—a monolithic door with a flat face—and frame and panel. There are lots of variations of these basic forms. Forms+ Surfaces makes some of the most intricately contoured wood doors I've seen (right photo, p. 10). (Addresses and phone numbers for companies mentioned are listed on p. 13.)

In general, the fancier the door, the more expensive it is. As you begin to step down in price, watch the moldings, the carvings and the contours disappear. And watch stiles and rails get narrower, and panels get thinner. Nord's low-priced doors have ⁹⁄₁₆-in. and ¾-in. panels; their more-expensive doors have 1⅜-in. panels.

Dealing with wood movement—Most people prefer the look of wood to steel and fiberglass. Plus, wood is easy to work; you can plane it, trim it, mortise it, add a dead bolt; and if you scratch or gouge a wood door, you can fill it, sand it and refinish it. On the other hand, wood warps, splits, swells, rots and has a low R-value. But wood-door manufacturers are doing something about these shortcomings.

Most mass-produced wood doors are veneer over an engineered-wood core (bottom left photo, facing page). The cores are either finger-jointed solid stock, such as fir or hemlock, or an exterior-grade fiberboard core.

My only concern about such doors is that the veneers are thin: ¹⁄₁₆ in. on a Simpson Mastermark, and that's before factory sanding. Pease's Registry II line has engineered-wood panels with membrane veneer, which gets applied under steam and pressure. The veneer is ¹⁄₄₂ in. thick. It wouldn't take much to gouge such thin veneer and give moisture a place to enter.

However, Gary Katz, a carpenter in Reseda, California, who's been working with doors for 20 years, figures that less than 3% of the laminated doors he's come across had delaminated. And Matt Holmstrom, a carpenter in Bedford, Virginia, with even more experience under his belt, remembers seeing only one front door with delaminated veneer.

Many door manufacturers guarantee that their panels will not split. These manufacturers glue two half-thickness panels together with the grain turned so that the panels won't cup in the same direction (bottom left photo, below). At least two high-end door manufacturers—A. A. W., Inc., and Lamson-Taylor—use this technology in their stiles. A built-up stile eliminates splitting and minimizes warping. Feather River Doors feature five-ply engineered stiles (bottom right photo, below) for

Wood door, foam core. Lamson-Taylor doors have several innovative features: A two-piece stile and rail minimizes wood movement; a ½-in. polyisocyanate core provides R-5.

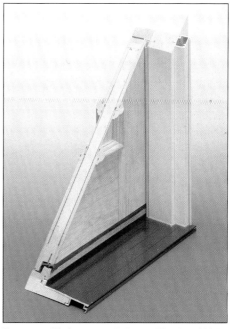

Energy-efficient wood door. Weather Shield Signature Series doors have cross-banded oak veneer glued to an insulated steel door. A prehung unit costs about $2,000.

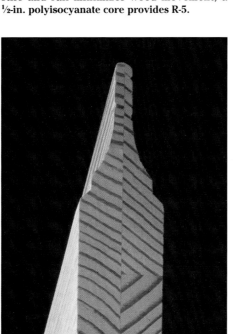

Nonsplitting panel. Simpson's Mastermark doors have ⅞-in. or 1¼-in. thick Innerbond panels. Because the panels are sliced and laminated, Simpson guarantees they won't split.

Nonsplitting stile. To avoid problems like splitting and warping, Feather River Doors builds doors with five-piece stiles. The core is vertical-grain fir; the outer layers are mahogany.

Top left and bottom right photos: Sloan Howard

greater stability. The core is vertical-grain fir; the edges and the faces are the same species as the door panels.

Insulated wood doors—One of the problems with a standard 1¾-in. thick wood entry door is its low R-value: R-2 at best. Panels often are thinner (Morgan's panels, for instance, are 1⅛ in.), making the whole door closer to R-1. To improve energy efficiency, several manufacturers have come up with ways to insulate wood doors. Weather Shield's Signature Series (top right photo, p. 9) has cross-banded oak veneer facings (the same process used in making plywood) glued with exterior adhesive to an insulated steel door. The door has a lifetime warranty on its stiles, rails and panels, which are guaranteed not to split as long as the door is finished immediately. Signature Series rates about R-8.

Lamson-Taylor builds wood doors with ½-in. polyisocyanate cores (left photo, below, and top left photo, p. 9). Their doors weigh in with an R-value of about 5.

Finishing and maintaining wood—Most wood-door manufacturers offer one- or two-year warranties. But those warranties won't be honored if you don't finish and maintain the door properly, which means following the manufacturer's instructions at the back of its brochure. Some manufacturers—Pease and IWP are two—offer a prefinished door.

If you finish the door yourself, hang it first to be sure it fits, then take it down and remove all hardware. Seal all six edges. If you plan to paint the door (these doors are graded "standard"), don't use lacquer products because they're not meant for exterior use. If you're staining (these doors are graded "premium" or "select"), apply at least two top coats of UV-inhibiting polyurethane or spar varnish. Some finishes will cause dark stains to appear in oak, so you should experiment in an inconspicuous area like the hinge edge. Among the common softwoods, hemlock paints easier than fir because there's no resin; fir stains easily; and pine is the easiest to finish and holds its finish better than hemlock or fir.

Because dark colors soak up sunlight and cause the door to heat up, paint doors a light color to reduce chances of warping or veneer checking. If the door has clear glazing, make sure

An insulated oak door. The Round Top Door by Lamson-Taylor has a ½-in. polyisocyanate core in its 1¾-in. thick panels, making them the same width as the stiles and the rails. Price not available.

Contours that only wood can provide. This solid Honduran mahogany Series 1000 door by Forms+Surfaces features a Gothic arch with R-Squared Linear side panels. Prices range from $1,300 to $2,000.

you coat the glazing putty (with paint or varnish) to protect against water leaks.

Maintenance involves monitoring for hairline cracks, textural changes, color changes or milkiness in the top coat. These are signs that it's time for a new top coat. The best protection for a wood door is an overhang or canopy. Matt Holmstrom won't even install a wood door in an exposed area. If the opening's exposed to weather, he recommends a steel door.

Making steel look good—Durability, insulation, security, low maintenance, low cost—if these features are important, a steel door is the best choice. Steel surpasses wood in all of these areas. Unfortunately, most people feel the same as architect Sarah Susanka, who says, "Steel doors are ugly."

Steel-door manufacturers are doing their darndest to dispel this widely held belief. For example, Pease recently developed a die that profiles steel-door skins with a raised molding and an indented panel. Called the Ever-Straight High Profile (top left photo, p. 8), the door looks better than an embossed door stamped with an indented panel because the raised moldings create light and shadows.

Several companies offer simulated wood-grain textures. For about $35 extra, Castlegate offers steel skins pressed with patterns that resemble oak or mahogany. The problem is that all of the grain runs vertically, which looks pretty goofy on the rails, where grain should run horizontally (bottom photo, right).

For a more-authentic stile-and-rail grain pattern, some companies bake a 10-mil vinyl overlay on their steel doors (top photo, right). The simulated wood grain runs vertically in the stiles and horizontally in the rails. All of these wood-grain doors—steel or vinyl—are meant to be finished with heavily pigmented stains and clear UV-inhibiting top coats.

A foam-core sandwich—Most steel doors consist of either a wood or steel frame with a pair of metal skins glued over an insulating-foam core. This core is either molded expanded polystyrene (MEPS) or polyurethane. You'll know if the core is polyurethane because the door maker will trumpet an R-value of 15 or so. The fanfare subsides when it comes to telling you that polyurethane's R-value diminishes over time, giving urethane-core doors a stabilized R-value of around R-10.5. As its R-value declines, polyurethane also releases pollutants claimed to cause ozone depletion.

The other insulating core, MEPS, doesn't have the high initial R-value (about R-4 per in.; a 1¾-in. door is about R-6), but it retains its R-value and doesn't deplete ozone.

If you live in a cold climate, make sure there's an insulating thermal break between the steel skins (drawings, p. 12); otherside, frost can form inside the door. Many steel doors have wood edges that provide a thermal break, but some doors have steel edges with vinyl thermal breaks. Taylor doors are steel-edged and appear to have no thermal break. The Taylor representative that I spoke with claimed that the adhesive bonding

Vinyl-coated steel. This steel door, the Benchmark Legend, has wood-grained texture, 10-mil PVC vinyl baked on. The coating is guaranteed for 10 years against delamination. Legend looks and finishes similar to wood, but its R-value is 10.5. The prehung unit without the sidelite costs about $350.

The grain runs vertically. Both steel and fiberglass doors try hard to look like wood. Some are more successful than others. Castlegate stamps its steel skins with a wood-grain texture. The skin is treated to be finished like wood, but it won't fool too many people.

Steel-door edges. *Many steel doors have wood frames, some of which can be planed to fit in an existing opening. Castlegate claims a steel-edge door is more secure. But if there's no thermal break between the exterior metal face and the interior metal face, condensation and frost will develop on the door's interior face. Also, steel-edge doors don't have mortised hinges.*

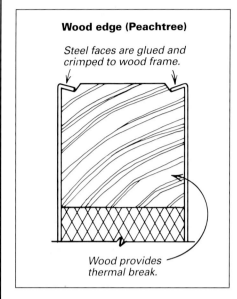

Wood edge (Peachtree)
Steel faces are glued and crimped to wood frame.
Wood provides thermal break.

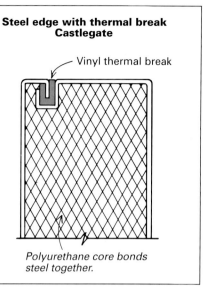

Steel edge with thermal break Castlegate
Vinyl thermal break
Polyurethane core bonds steel together.

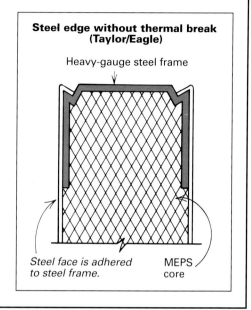

Steel edge without thermal break (Taylor/Eagle)
Heavy-gauge steel frame
Steel face is adhered to steel frame.
MEPS core

the door face to the steel edge serves as the thermal break.

Especially for remodeling, Gary Katz prefers steel doors with wood edges that project beyond the steel skins about ¼ in. These doors can be hung in an out-of-plumb or unsquare jamb successfully because you can plane the edges and mount the hardware in wood.

Finishing steel doors—Steel doors are sold with a primer coat; it's up to the consumer to finish them. Taylor and Stanley recommend latex paint, and Pease recommends water-based acrylic paint. Manufacturers require that a door be finished within 40 to 45 days after installation. Stanley prohibits dark colors on doors with plastic decorative trim, called plants, or glazing beads, because these elements will distort when exposed to direct sunlight. Pease tells you to not paint the seam between the plants and the door. Good luck. Pease also states that using an improperly ventilated storm door may void the door's warranty. On all doors, avoid painting the weatherstripping and the sweep.

As with any door, how often you must repaint a steel door depends on its exposure. A door that basks in the sun day after day will require repainting a lot more frequently than a door that's protected by a canopy. If your door sits in the sunshine, you can reduce maintenance by painting it a light color.

Any carpenter who's pushed open a steel door with a sheet of plywood knows that steel doors dent. One carpenter I talked to puts cardboard over steel doors on an active job site, but he says it's almost impossible to keep the doors dent-free. If a steel door is dented, it can be repaired with auto-body filler from a hardware store.

Fiberglass—Therma-Tru introduced the first fiberglass entry door in the early 1980s. It had the

Comparing door prices

Although there's a lot of overlap, steel doors are generally the least expensive, followed by fiberglass and then wood. There's a lot more to an exterior door than its price. But to give you an idea of how steel, fiberglass and wood stack up against each other price-wise, I called three local lumberyards, then averaged their prices for a solid (no glazing), six-panel, stile-and-rail 3-ft. by 6-ft. 8-in. by 1¾-in. door slab—a wood door, a steel door and a fiberglass door.

Wood—I asked each lumberyard for a price on its most popular door. A solid-pine door with 1⅛-in. panels topped the charts. Morgan M-7100: $608.

Steel—I asked for prices on a door with 24-ga. skins, wood edges and a polyurethane core because this was the combination seen the most in door manufacturers' catalogs. Brosco BE 70 (Perma-Door): $108.83.

Fiberglass—I got a price on the original Therma-Tru Fiber Classic, not the new composite door. Therma-Tru Fiber Classic FC 60: $211. *—R. Z.*

R-value of a steel door but didn't dent like one; it looked an awful lot like an oak door but was priced less. Now Pease, Peachtree, Castlegate and Stanley are among the companies that offer fiberglass doors, but there aren't many styles or sizes available. For example, there are no crossbuck designs, and there's only one height: 6 ft. 8 in. Glazing options between steel and fiberglass are comparable. Fiberglass sidelites and transoms are available, too.

Cost-conscious builders often go for a steel door, but a fiberglass door might make a house easier to sell. A front door is a home's focal point and your first contact with the home. Fiberglass approaches the warmth of wood but with lower maintenance. And if a door will sit around unfinished at a job site, fiberglass won't get beat up like steel and won't warp the way wood does.

John Loftus, a custom builder in Albany, New York, has installed several fiberglass doors. In fact, the cover of *FHB* #63 (right photo, p. 8) features a Colonial entrance built by Loftus, and it's got a Therma-Tru Fiber Classic fiberglass door. Even though fiberglass doors are more expensive than steel, Loftus likes the wood-grain look and the natural finish. He also considers fiberglass a good choice for doors that will be exposed to the weather.

Most fiberglass-door brochures tout the fact that these doors won't crack, split, warp, check or rot. But according to David Pease, III, of Pease Industries, which manufactures wood, steel and fiberglass doors, fiberglass has the highest coefficient of expansion, followed by steel. Wood has the lowest. So a fiberglass door will experience the most thermal bowing. If a door bows a lot, it can lose contact with the weatherstripping, allowing air and water to infiltrate. I asked Loftus about thermal bow, and he said that no one for whom he's installed a fiberglass door has complained about infiltration problems.

Improving on fiberglass—Therma-Tru's new door, the Classic Craft, looks more like wood than its Fiber Classic door, thanks to a new type of fiberglass composite. The Classic Craft is heavier than the Fiber Classic because of the density of the composite and the door's solid-oak frame.

Peachtree Doors also offers a new fiberglass-composite door, called the Newport. Both Therma-Tru and Peachtree claim that fiberglass composite absorbs finish better and holds a finish longer than the original fiberglass doors. Neither company would say exactly what makes up the fiberglass composite.

All fiberglass doors are constructed just like steel doors with wood edges and contain the same insulating cores. Unlike steel doors, which have slightly rounded edges, fiberglass doors have square edges, and the exposed wood frame, or edge, is usually veneered with oak.

Even though fiberglass doors have been on the market less than 15 years, manufacturers typically offer longer warranties on fiberglass doors than on steel: Castlegate offers a 30-year warranty on its fiberglass doors and a 20-year warranty on its steel doors; Stanley offers a lifetime limited warranty on fiberglass and a five-year limited warranty on steel.

Installation and maintenance—I called the Home Depot up the street to find out if they stocked fiberglass doors. They don't, but I could order one: Stanley's Prodigy. It takes a week to 10 days to get it. Loftus buys prehung Therma-Tru fiberglass doors through his lumberyard. The lead time is good: about a week or two from order to delivery. And he says installing a prehung fiberglass door is no different from installing a wood door: shim and nail through the jamb. If you want to add a dead bolt, Loftus recommends making the cross bore with a Forstner bit.

"The biggest pain was in finishing the door," Loftus told me, because it took a while for the UV-inhibiting top coat to dry. He followed the manufacturer's instructions and had no problem. You have to wipe the door with mineral spirits first and allow it to dry; "Otherwise, it comes out blotchy," Loftus said.

You can get fiberglass doors factory finished; it costs about $100 extra. If you finish the door yourself, fiberglass finishing kits are available from door manufacturers; Therma-Tru's lists for $19.95. The kits contain tubes of oil-based paint, a UV-inhibiting spray-on top coat and sample pieces of fiberglass on which you can practice. If you don't like what you see, you can remove the stain with mineral spirits and start over. It's also possible to finish the door with a solid-color, linseed-oil base stain. Don't sand fiberglass doors because it wears away the wood-grain texture.

To maintain a fiberglass door, you simply reapply the finish coat every two or three years. Of course, the time between refinishing depends on the door's exposure to the weather. Nicks or scratches can be repaired with fiberglass paste available from a hardware store. □

Rich Ziegner is an assistant editor at Fine Homebuilding. *Photos courtesy of manufacturers except where noted.*

Door manufacturers

These are the addresses and phone numbers of the door manufacturers I mentioned in the article. Many more door manufacturers exist.

A. A. W., Inc.
14201 South Main St.
Los Angeles, Calif. 90061
(310) 516-1089
Custom wood entry doors.

Benchmark
Box 7387
Fredericksburg, Va. 22404
(703) 898-5700
Steel entry doors.

Castlegate
911 East Jefferson St.
Pittsburg, Kan. 66762
(800) 835-0364
Steel and fiberglass entry doors.

Feather River Door
P. O. Box 477
Durham, Calif. 95938
(800) 284-DOOR
Custom wood entry doors.

Forms+Surfaces
Box 5215,
Santa Barbara, Calif. 93150
(805) 684-8626
Wood and clad-wood entry doors.

Jeld-Wen (Including Challenge, IWP and Nord)
3303 Lakeport Blvd.
Klamath Falls, Ore. 97601
(800) 877-9482
Wood, steel and fiberglass entry doors.

Lamson-Taylor Doors
Tucker Road
South Acworth, N. H. 03607
(603) 835-2992
Custom insulated wood entry doors.

Morgan Manufacturing
601 Oregon St.
Oshkosh, Wis. 54903
(800) 766-1992
Wood entry doors.

Peachtree Doors, Inc.
4350 Peachtree Industrial Blvd.
Norcross, Ga. 30071
(404) 497-2000
Steel and fiberglass entry doors.

Pease Industries
7100 Dixie Highway
Fairfield, Ohio 45014-8001
(800) 543-1180
Wood, steel and fiberglass entry doors.

Perma-Door/Taylor Building Products (including Eagle)
631 N. 1st St.
P. O. Box 457
West Branch, Mich. 48661
(800) 842-3667
Steel entry doors.

Simpson Door Co.
Box 210
McCleary, Wash. 98557
(800) 952-4057
Wood entry doors.

Stanley Door Systems
1225 East Maple Road
Troy, Mich. 48083
(810) 528-1400
Steel and fiberglass entry doors.

Therma-Tru Corp.
P. O. Box 8780
Maumee, Ohio 43537
(800) 537-8827
Steel and fiberglass entry doors.

Weather Shield Manufacturing, Inc.
1 Weather Shield Plaza
Medford, Wis. 54451
(715) 748-2100
Steel and insulated wood entry doors.

Door associations

Insulated Steel Door Institute
30200 Detroit Road
Cleveland, Ohio 44145-1967
(216) 899-0010
Trade organization representing manufacturers of steel doors.

National Wood Window and Door Association
1400 East Touhy Ave., Suite G-54
Des Plaines, Ill. 60018
(708) 299-5200
Trade organization representing manufacturers of wood doors.

Connecticut River Valley Entrance
Reproducing a famous 18th-century doorway

by Gregory Schipa

This past spring we were moving into the last stages of the restoration of the Burley house, a beautiful Georgian structure built in Newmarket, N.H., around 1710. It had been dismantled, moved to Riverton, Vt., and re-erected on a new site.

Although we had found a Federal entrance on the building at its original site, framing mortises in the structure left no doubt that the house had once boasted double, 24-in. wide doors. As part of the restoration, the owner wanted us to recreate a Connecticut River Valley broken-pediment entrance. Since these beautiful entrances became fashionable around 1750 in the Newmarket area, she surmised that the old Burley house might have gotten one as part of a facelift to reflect the family's increasing affluence. The proportions of the building were right, and I was easily persuaded. After all, a job like this is a joiner's dream. We decided that it would be best to join such an entry by hand in the original manner. We had only to do a little research and then find our favorite example.

Development of the style—In 1636, settlers of the Massachusetts Bay Colony began to move westward from Boston to the Connecticut River Valley. Preferring the isolation and dangers of the wilderness to the overcrowding and religious intolerance in the Boston area, they first settled Springfield, Mass., and then Hartford, Windsor and Wethersfield, Conn. The migration continued for another 100 years, extending up the Connecticut River to New Hampshire and Vermont. It was not long before these pioneers began adding distinctive touches to the houses they built.

Most of the early work in the Colonies had been Jacobean in style, and was executed by English craftsmen, but the Connecticut River Valley towns soon developed their own unique details and embellishments, while staying with the same basic house plan. With the start of the Georgian Period, change accelerated. It can most easily be traced in the treatment of the front entrance, which rapidly evolved from a simply framed batten door through embellishment with the common architrave, transom and crown to the introduction (about 1700) of pilasters, double doors and the classic three horizontal members: architrave, frieze and cornice. Connecticut River Valley craftsmen lavished ever-increasing creativity on these entrances, apparently in celebration of the lessening dangers and growing wealth and freedom of the period. Soon cornice members evolved into full pediments, and before the middle of the 18th century, these elaborate pediments were further embellished by being broken and ended with volutes or rosettes.

The historic Burley house was moved from its original location to central Vermont, and its new entry was modeled after the one on the Rev. John Williams house in Deerfield, Mass.

Finding our model—Some of the very best examples of the Connecticut River Valley entrance can be found at Old Deerfield Village, Mass. Perhaps the best known of these is on the Rev. John Williams house (see p. 17). We chose to reproduce this particular entrance because its door opening was the same size as the Burley house rough opening. Also, the dimensions of the two houses were almost the same, and the window placements were similar. The Williams entrance also seemed to represent the best of the Connecticut River Valley pilastered doorways, with an early Jacobean feeling reflected in its steep, handcarved moldings (very few of which are classical), and its high raised-panel tombstone pedestals. Its massive broken-scroll pediment is very steep and abrupt, and its double pilasters are deeply fluted, with carved rosettes in the necking. It is perfectly balanced and well detailed. We hoped we could match the work of the 18th-century craftsman who built it.

Getting started—My foreman, Richard Tintle, and I first went to Deerfield to measure, draw and photograph the entryway. We figured it would take the two of us at least four weeks to do the job. We built the entrance primarily of 5/4 and ¾-in. clear local white pine. The stock had to be clear because we were cutting all moldings by hand. We could usually get two moldings from each 1x4 board, but cost was very much a factor, so we tried to use every extra strip of pine for the smaller moldings.

Very little was necessary to prepare the building to receive the work. The size of the rough opening was already dictated by the double stud mortises in the frame, and at 51 in. wide by 80 in. high, it was almost exactly the same as the one in the Williams house. The windows in the Burley house were far apart, and we had sheathed with ⅝-in. plywood, so we had a large, flat, blank space on which we drew the outline of the entire entrance. This gave us an immediate feel for its impressive dimensions, and made it easy for us to figure how much wood we would need for our first step, the underlayment. We also rigged a tarp so the entry would be under cover until it was completed and primed.

The underlayment, more properly called the clapboard catch, consists of the boards to which the entire built-up entrance detail is attached. We used two 8¾-in. wide boards of nominal 1-in. stock on each side, planning to hide the seam under the pilasters. Two horizontal 11½-in. wide boards across the top extended above the eventual level of the drip cap (the horizontal member of the cornice), and high enough to accept the bottom of the broken-scroll pediment. Two more 1x12s were later used to back the pediment and match its curve. All the boards used for the clapboard catch were joined by hand and jackplaned. The Williams entrance had been further embellished with quoin work (carved V-grooves imitating masonry joints). On our project, we carved these grooves with a V-gouge.

Pedestals and pilasters—Once we were done with the clapboard catch, we began working from the ground up. The base of a pilaster is the plinth block. Because it is close to the moist ground, it is usually the first part of an entrance to rot out. This had happened on the Williams house, where the original plinth blocks had been replaced very shabbily with old rough-cut studs. We used solid sugar-pine blocks on the Burley house, hand-dressed to 7 in. high, 12 in. wide and 2¾ in. deep.

The pedestals for the pilasters rise immediately from the plinth blocks. They are doubled, the under-pedestal being 2½ in. wider than the one on top. They both called for carved quoin work, and the surface pedestal on each side is a tombstone panel (photo right). We were surprised to observe that on the Williams house these little panels were not joined in the customary manner, with stiles and rails, but were actually carved 1-in. boards. We followed suit on our bench, using butt chisels and a V-gouge.

The double pilasters themselves represented quite a challenge. Both the pilasters and under-pilasters were fluted, and their entasis, or taper, had to be reflected in the fluting. To do this, we gradually raised the plane to reduce the depth of each flute toward the top. This also had the effect of reducing its width of cut. We found the Stanley #55 multiplane (photo below right) handy for this job, but its fence could not be used. We had to fabricate one that would follow the taper. The final 16 in. or so of each flute had to be carved with a U-gouge after every few passes, or the plane would ride up. I also clamped a small tab of sheet metal under the stop-block, extending from it to the end of each flute. This protected the flat area beyond the flutes from being scarred as the plane's bottom passed over it.

The fluted part of the pilasters is about 56 in. long, but the pilasters themselves extend to about 77 in., past the capital molding group and carved necking, under the architrave and frieze, finally ending at the bottom of the cornice, where the molding of dentiled corona steps around its top. The pilasters taper from $7\frac{7}{16}$ in. at the bottom to 6½ in. at the top. The under-pilasters are proportionately wider, reflecting the same shadow-line as the pedestals. Pilaster and under-pilaster each project $1\frac{1}{16}$ in., and they are both fluted on their sides as well as on their fronts. Before finishing the pilasters, we laid out and carved the 6-in. dia., six-petal rosette on the necking of each. When we finished this part of the project, we had already worked 70 man-hours.

The moldings of the pedestal cap and those of the pilaster base form a single, solid group. They are quite primitive, mostly a series of hollows and rounds—early Jacobean rather than classical. Nevertheless, they have considerable impact because of the depth to which they're cut and the distance they project. We cut out these and other moldings with antique planes (photo top right) and the Stanley #55. We found

The tombstone panel of one of the pilaster pedestals is nailed up, below. The under-pedestal extends 1¼ in. out from each side of the face panel, and both sit on the solid sugar-pine plinth block. Two 8¾-in. boards make up the underlayment, or clapboard catch, to which everything else is attached. All the elements have been gouged to resemble stonework.

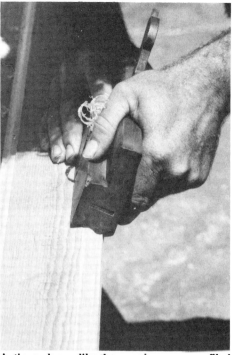

Antique planes, like the one above, are profiled to shape a specific molding. They came in handy on this job, where so many traditional shapes were required. A Stanley #55 plane (below) was used to cut the pilasters' fluting and, as shown here, to shape many of the entryway's moldings. This tool, designed for joiners, accepts blades of many different types and patterns.

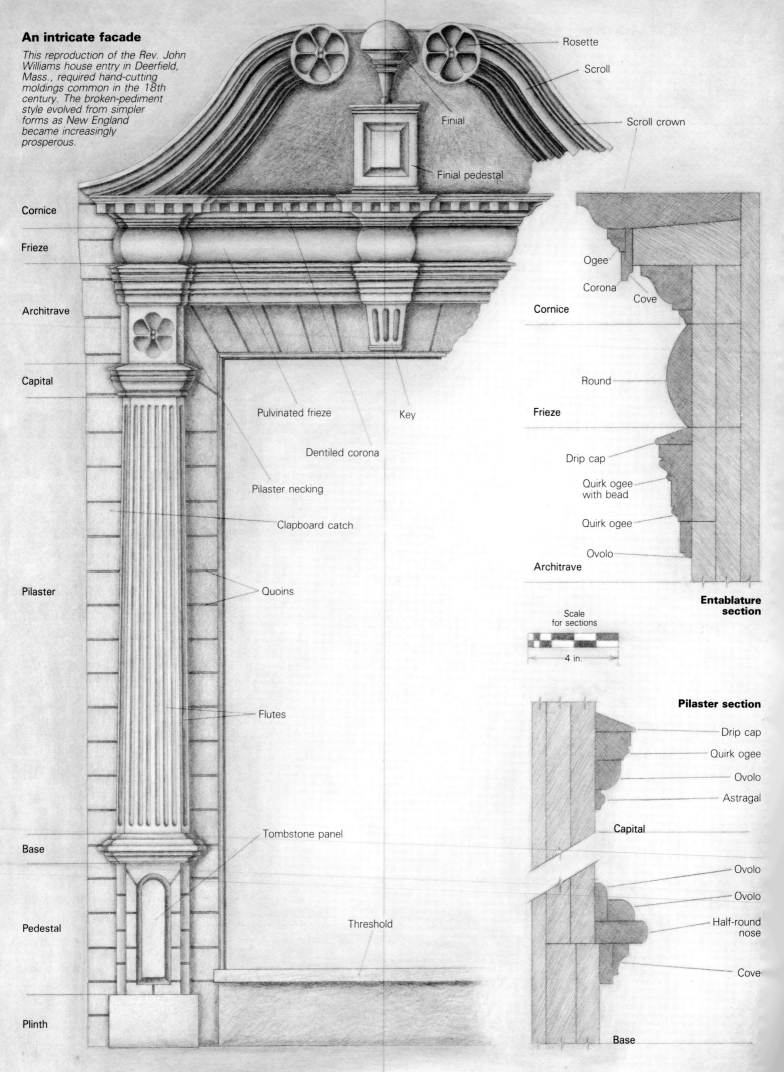

The perilous history of the Rev. John Williams and his house

In histories of Deerfield, Mass., the Rev. John Williams is known as "the redeemed captive." His house was burned to the ground and several of his children were killed during the French-fomented Indian raid of 1704, and Williams was carried off a prisoner to Canada. His wife was murdered on the march, and his 10-year-old daughter Eunice was adopted by the Caughnawagas. She converted to Catholicism and eventually married an Indian named Amrusus—a process that one partisan chronicler called "her lapse into barbarism." In 1706, Williams himself was exchanged for a Frenchman held prisoner by the English. His loyal congregation enticed him back to the wilderness from the safety of Boston by building him a new house in 1707.

When John Williams died, the house passed first to his second wife, and then, upon her death in the 1750s, to their son Elijah. A subsequent owner of Elijah's house bequeathed the structure to Deerfield Academy in 1875.

In the late 19th century, old houses were often destroyed with little regard for their historic value, and the school planned to raze the house to make room for a new building. But George Sheldon, president of the Pocumtuck Valley Association, saved the Williams house—and its glorious entry—by mounting a campaign that linked it closely with the Rev. Mr. Williams, the most romantic figure in the town's history. As researchers Amelia F. Miller and Donald L. Bunce put it, "With skillful omissions and careful wording, he allowed his readers to believe that this house was the one built in 1707." It wasn't. Miller and Bunce say that the 1707 house was pulled down, and a new house built in 1759 or 1760. Elijah had become a man of means and wanted a grander residence.

Miller and Bunce also found evidence that the great broken-pediment entryway was built by Lieutenant Samuel Partridge in the summer of 1760. Partridge had been with General Wolfe at Quebec, and was related to Elijah by marriage. He must also have been a master of his trade who could demand top dollar. Elijah's account book for September shows a disbursement of £39/0-2 (39 pounds, two pence) in an era when journeyman carpenters were making only a few shillings a day.
—*Mark Alvarez*

that applying molding around a double pedestal and pilaster requires three times the miter work as on single elements.

The pilaster capital molding group of the Williams entrance lines up just below the head jamb. It is steep and simple, which is appropriate to the period, but its moldings are more recognizably classic, with a ⅞-in. astragal, a ¼-in. ovolo and a ⅝-in. ogee. The group has its own beveled drip cap, cut from a ¾-in. board and shaped with a bench plane.

The architrave—We thought we had done a few miters up to this point, but as we worked on the 95 pieces of the architrave band of moldings, the word began to take on a new meaning. The band has five components (including a spacer), and it steps around both double pilasters and the tapered key. The key, with its own handcarved flutes, extends from the door head to the bottom of the cornice, and matches the size and taper of the quoins. The architrave band is 5 9/16 in. wide, made up of a 1 9/16-in. bottom piece cut to a ¼-in. ovolo molding, followed by a 1¾-in. piece cut to a ⅝-in. ogee, and then a 1½-in. ogee with a bead. The full band has its own ¾-in. tapered drip cap, like the ones on the pilaster capitals. The architrave is ½ in. deep at its bottom and 1⅝ in. at its top—considerably less projection than the pilaster base and cap moldings.

Immediately above the architrave, and sitting flush upon its drip cap, the craftsman at the Williams house had fashioned a rounded, or pulvinated, frieze. We found that we could fabricate this element with a large hollow, a tool that was specifically made for the pulvinated molding, or with a bench plane, scratch-stock and sandpaper. The inaccuracy inherent in the handmade process could easily be worked out with a little additional sanding after the frieze had been mitered and fitted.

Sitting upon the frieze of the Williams entrance, and fitted against the bottom of the cornice like a bed molding, we found the classic 18th-century crown molding. Consisting of a small ogee on the bottom, with a bead above and a large reverse ogee on top, this molding

The mitered moldings of the entry required painstaking sanding. At left, Richard Tintle works on the crown molding beneath the cornice. The clapboard catch, grooved to resemble quoins, is visible to the left of the pilaster.

The scrolls (bottom left) were glued up from sugar-pine planks, and the curves were smoothed with an antique compass plane. On the facing page, the completed pediment.

The finial between the scrolls sits on a paneled pedestal, built as a sturdy box. The side panels are flat, but the front panel, below, is carved. The box will get a hipped top.

(photo above left) was one of 18th-century New England's most consistently popular architectural embellishments. We nailed it in place before we put up the frieze. We were very lucky to own the appropriate antique plane, because its use was the key to cornice and pediment detail on this particular entrance. The crown molding is repeated on the broken scroll, where it is mitered into the small ogee on the cornice fascia.

The pediment—A hundred and twelve man-hours after leaving to measure the famous entrance in Old Deerfield, we were ready to start the pediment. We had made and fitted 270 pieces up to this point. But the craftsman who worked on the original had saved his most creative efforts for last. The massive cornice, rising in steep reverse curves to form the broken pediment, and terminating in bold carved rosettes, is one of the most impressive details of 18th-century domestic architecture. We knew it would be a challenge to reproduce.

The cornice structure begins with the usual horizontal element, which serves as the drip cap for the entrance below. On our entry, we reproduced this with a 2-in. thick, 5-in. wide sugar-pine plank, which we beveled with a jack plane to shed water. We added 1-in. blocks to simulate the step-outs of the pilasters and the key below. The fascia applied to this plank has a $5/16$-in. reveal on the bottom, and a small $1/4$-in. cove cut into its back below that. Its front bottom, also called the corona, has small dentil-

Jim Eaton

like cuts, 1¼ in. square. These can be cut with a fillister plane, or a rabbet plane with a fence clamped to it. It is then topped with a ⅝-in. ogee molding, which we mitered into the bottom ogee of the crown molding, on the return of the broken-scroll pediment to the house.

We laid out the scrolls using a string and a pencil on a very large table. This was hard because we had only our photographs as guides, and we wanted to match the proportions of the original builders. To miss here would sacrifice a lot of very careful work. Once we felt we had proper proportions, we drew out templates on plywood, including all of the cove and thumbnail bed moldings that would follow the reverse curve pediment immediately below it. Then, as we cut them off, each snake-like strip became a different template. The scrolls themselves we cut from laminated sugar-pine planks. We used a beautiful old compass plane to smooth them after an initial rough cutting (photo facing page, bottom). On the rough scroll we attached ½-in. fascia (again with the reveal and cove on the bottom), but here we discovered a real quirk of the craftsman who built the Williams entry. He had used a different crown on the scroll than on its returns to the house. Although he used a large reverse ogee with a bead and a small cove as his crown molding cut on the scroll, he used the very typical large reverse ogee, bead and small ogee for the returns. The large reverse ogees were the same size, however, and were mitered at their meeting. The small ogee was mitered into the ogee above the corona. It was a tricky resolution. We made the curved moldings by simply carving them and then working them hard with scratch-stock and sandpaper. We used the large reverse curve of our old crown molding plane as the template for the profile of the scroll crown.

The resolution—Large carved rosettes resolved the scroll tops. They were very similar to the rosettes on the pilaster necking, but were 2 in. larger, and our carving was in bas relief. We did all the rosette carving with U-shaped gouges. The pediment, from clapboard catch to the outside edge of the crown molding, is the same 8-in. depth as the block the rosettes are carved in, so we could terminate all moldings on the circular barrel of the rosette.

Between the rosettes, the pediment on the Williams house had been further embellished with a turned finial on a carved, paneled pedestal. Here we again struggled with our layout, for the finial had been too high on the real entrance for us to measure. Our drawing had to be our template, and when we finally settled on the proportions, we turned it out easily on the lathe—just as it had been done originally, if with a different power source. We left a large round tenon to fit into a mortise in the roof of the pedestal. The pedestal itself we made by first creating a box 8 in. wide by 12 in. high by 4 in. deep (small photo, facing page). Then we hand-carved a raised panel on its front, and flat panels on its sides (another oddity of the Williams entry). We fashioned a hipped top, flat only on the small portion meeting the bottom of the finial. A deep mortise accepted the finial's substantial tenon. We flashed the cornice, then set the pedestal and finial in place.

So there we were. Three-hundred and fourteen handcarved pieces, 154 man-hours of labor, and we had ourselves an entrance—and a lot of respect for those Connecticut River Valley craftsmen. In hopes of its lasting half as long as the original, we used handmade rosehead nails where they would show, and galvanized nails everywhere else. We also used almost two quarts of glue and untold pounds of lead for flashing. In addition to the horizontal cornice, we also flashed the scroll pediment, overlapping the edge of the crown molding by almost ¼ in. We even fitted the finial with a small leaden skull-cap.

After the awesome entrance, we built the doors, almost as an afterthought. They were real beauties, though. The client chose as our model a fine set of early four-paneled doors, again from Old Deerfield. They were fitted with beaded vertical battens on the inside, large strap hinges, and a huge bar and staples lock. The battens were fastened with clinched roseheads, the stiles and rails with pegs. The client was lucky enough still to have the huge granite step from the original house. We had only to move it gently up under the plinth blocks to make the job complete. □

Gregory Schipa is president of Weather Hill Restoration Co. in Waitsfield, Vt.

A Greek Revival Addition
Joining stock and custom-milled trim to make a formal entryway

by Joseph Beals

A modern and unremarkable Cape Cod cottage needed a new gable-end door and entry. The existing frame was falling apart, the laminated oak sill had delaminated, the brick stoop was crumbling and the apron under the door was rotting, along with the sidewall shingles that the stoop abutted. Because the entry had failed in so many ways, the owner took advantage of circumstance and asked us for a new entry. We chose to use two doors with an airlock to make an enclosed entrance.

Greek Revival architecture was the style of choice; not only is it common in this part of New England, but it also lends itself to full expression even on a small scale. The entryway had to be big enough to include the many formal elements that make up the Greek Revival style, but not so big as to overpower the gable end of the house.

In the transition from the marble buildings of ancient Greece to the woodwork of the early United States, Greek Revival architecture has retained the elements of form, yet has allowed a freedom of ornamentation ranging from the austere to the exquisitely ornate. A gable wall rather than a sidewall most often serves as the front of a Greek Revival house and is the focus of the design. An entablature surmounts the sidewalls under the roof and continues horizontally across the gable, forming a pediment above. Where it appears on the sidewalls, the entablature is typically surmounted by a crown molding, which is mitered to raking crown moldings that meet at the peak of the gable. The face of the pediment is recessed behind the rake trim and entabulature, creating a strong play of light and shadow. In Greek architecture, the gable-end entablature is supported by columns around the entire building, but in many Greek Revival buildings, the columns are replaced by pilasters and detailed corner trim, which echo the effect of columns without the function.

Designing the entryway—The width of the entryway was critical because of tight space. The owner requested a 36-in. outside door flanked by sidelights that he had salvaged some years before. I drew a front elevation incorporating a pair of fluted pilasters outboard of the sidelights on the front corners, and another pair between the sidelights and the door. I sketched the details of the entablature above the door so that I could derive the height of the sidewall and figure the length of rafters and location of cuts. The entryway depth was less critical, but it was important not to make it protrude too far. An interior depth of 4 ft. would allow a comfortable margin for opening the outside door, and would also mean that a single sheet of drywall could finish each interior sidewall. An invaluable resource for me during design was *New England Doorways* by Samuel Chamberlain (Hastings House Publishers, N. Y., 1939; out of print).

The entryway would have a 12:12 pitch roof—steeper than a traditional Greek Revival roof, but identical to the pitch of the main roof. This ensured that the entryway would look like part of the house rather than like a contrived addition. The steep gable makes an unusually tall pediment, so to soften the height we added a half-round window above the cornice. This solution was more luck than genius: the window was the top of an old door that had arrived in a pile of junk from another job, and it begged to be saved. A few days in the shop was enough time for me to saw it free, then fabricate the curved casings and add a sill.

Foundation and steps—After the existing brick stair and landing were broken up and removed, we found that rot in the wall was more

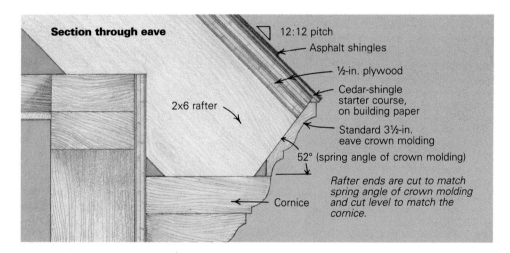

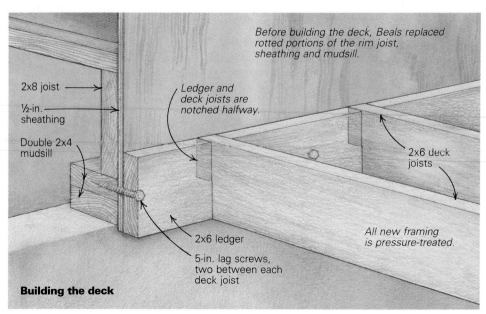

extensive than we had expected. Not only were the sill and rim joist compromised by water damage, they were also full of carpenter ants. Because a gable-end wall doesn't carry much load, replacing the rotten portions of the rim joist and sill and patching in new sheathing took only half a day.

The existing entry would remain in use during construction, so I built a foundation deck at once. To achieve the width we needed for door, pilasters and sidelights, the deck would be too wide to center on the existing door, as its left corner would lie over the water supply line. I moved the deck 2 ft. to the right to clear the water line, and later framed for a new door. Moving the door over solved a minor problem inside the house, where the entrance had been too close to the corner of the room to allow for usable space against the wall.

I lag-screwed a ledger to the new sill and fit notched pressure-treated 2x6 joists into notches cut in the ledger (bottom drawing, facing page). The 4x4 outer posts of the deck bear on cast-in-place concrete piers. The surface of the deck is 6½ in. below the floor of the house. The two lower steps have 12-in. treads, while the top step is 30 in. deep and serves as a landing (photo right). It also makes a graceful transition from the narrower steps to the door. In a traditional entry, the landing would be stone or brick and have an iron boot scraper fitted at one end, but those are costly details this project had to do without.

The cornice and pediment—The entry framing is conventional, but the roof required careful planning to accommodate the entablature. I cut the rafter tails to match the spring angle of the crown molding (the spring angle describes the cant of the molding in relation to the horizontal) and cut the bottoms level to provide support for the cornice (top drawing, facing page). To provide a solid backing for the crown molding at both rake and eave, I sheathed the roof deck with two layers of ½-in. plywood. At the gable face, I ripped the cantilevered ends of the plywood slightly to match the spring angle of the raking moldings.

The trim for the entryway was built up from standard lumberyard moldings and from moldings that I milled from 1x or 2x pine stock. The first trim I installed, after sheathing the entryway walls with plywood, was the top piece of the cornice, which I milled from 2x6 stock. I nailed it to the rafter bottoms on the sidewalls and toenailed it into the studs along the gable end. Along the gable end, where the top of the cornice is exposed to the weather, I planed its top to a 3° slope to allow for water runoff (drawing next page).

The pediment had to be finished next. In a small structure like this, the recessed face of the pediment is traditionally a monolithic surface—shingles or clapboards would destroy the purity of appearance. I built this one by edge-gluing lengths of clear pine with resorcinol adhesive, and sealed the joints with orange shellac to prevent glue lines from showing through the paint. Then I primed both sides with oil-base paint.

To attach the pediment, I worked lead flashing into the corner between the top of the cornice and the gable sheathing, then fastened the pediment to the studs with a few finish nails. I built up the rake trim—all but the crown molding—fitting it tight to the cornice at its lower ends, and mitered each piece at the peak (top drawing facing page). The next step was to fit the eave and raking crown moldings that are such a strong feature of Greek Revival architecture.

Matching raking to eave molding—On the Greek Revival houses of New England, eave crown moldings mitered to raking crown moldings are such a common sight that I was not prepared for the complexity of the joint between the two. A raking molding, which

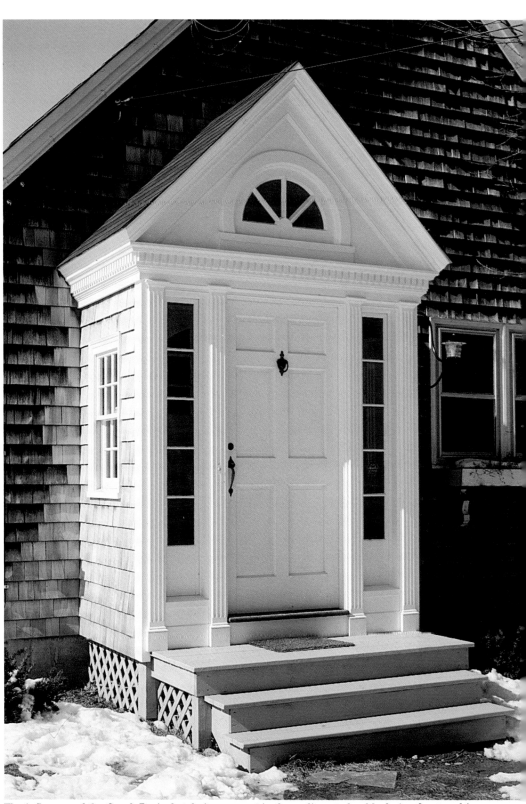

The influence of the Greek Revival style is apparent in the pedimented gable facing front and in the strong lines of the entablature of this entryway addition. The most challenging task to building a pedimented gable is mitering eave and raking crown moldings. Here, the author milled and hand-planed his own raking molding to a modified profile of the eave molding.

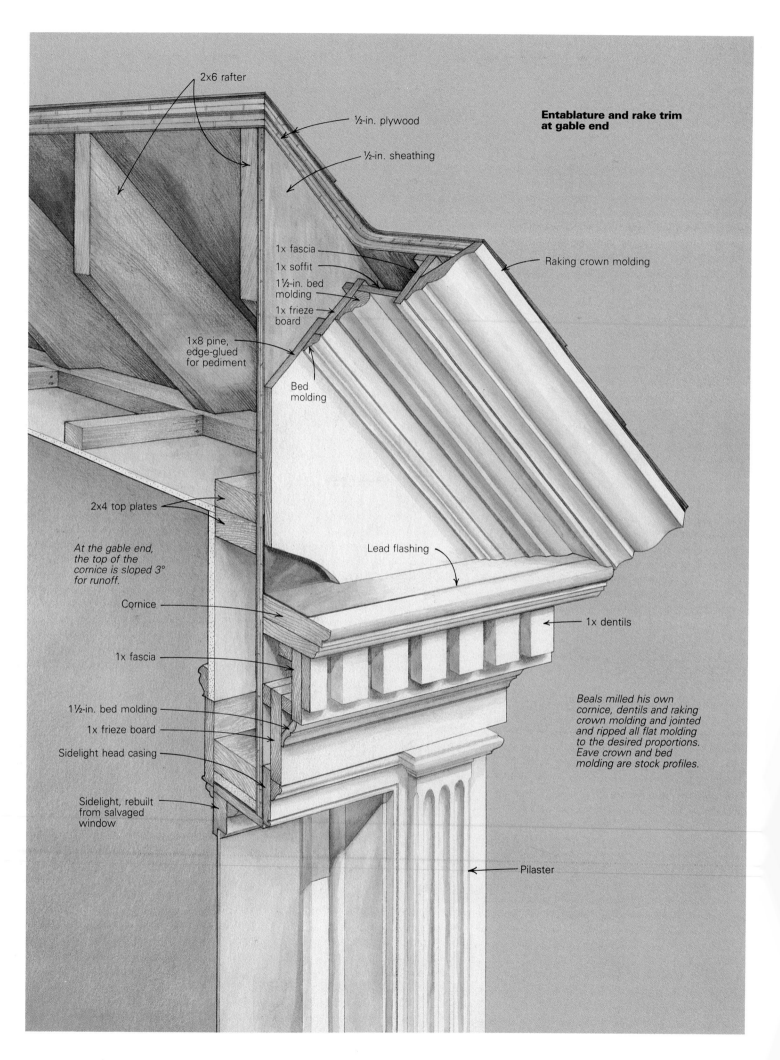

slopes downward and outward, can only meet an eave molding if its profile is milled wider than and in a precise distortion of the profile of the eave molding. The profile of the raking molding is a function of roof pitch, reaching a maximum elongation when the roof pitch is 1:1. A hundred years ago, it was doubtless common practice for lumberyards to stock suitable complements of eave and raking moldings. But those days are long past. Now we improvise.

I used standard lumberyard crown molding on the eave walls and mitered their outboard ends in the usual way, as if they were to intersect horizontally with another eave molding (drawing below). Then I held a 1x6 in place on the rake fascia in contact with the mitered end of the eave molding, and traced the eave profile onto the 1x6. To achieve the approximate profile of the raking molding, I gave a 1x6 a number of passes with different knives on the molding head of my table saw. I used molding planes and sandpaper to true the profile.

I marked the bottom miters of each length of raking molding with both a square and a bevel gauge, then cut proud of the line with a handsaw and trimmed with a block plane until the joint came up tight. This seems straightforward, but was really a devilish exercise. Intuition is no help: the miters of the raking moldings look improperly cut until the moment when the piece slips into position and the puzzle comes together (see *FHB* #41, pp. 64-65 for more on raking molding). My reference for this art of the past, as well as for other joinery techniques, is *Modern Carpentry* by Fred Hodgson, published in 1917. It's out of print, but a similar reference is George Ellis's *Modern Practical Joinery*, published in 1908 and reprinted by Linden Publishing Co. (3845 N. Blackstone, Fresno, Calif. 93726).

Completing the entablature—Before beginning work on the entablature, I closed the entryway to weather. I shingled the roof with asphalt shingles, beginning with a cedar-shingle starter course along each eave to keep a traditional appearance. (We did not consider the use of an aluminum drip edge, which is out of place in restoration or reproduction carpentry.) I replaced the wall shingles that I had stripped to accommodate the new entryway, except for a small space at the top of each sidewall where the entablature would butt against the house. I scribed and fitted those shingles after finishing the entablature. I nailed 5-in. corner boards to the sheathing, then built and installed sidewall windows and shingled the walls.

Making the components of the frieze was my next task. A number of visitors have looked in wonder at the dentil molding and remarked on the skill and patience required to fit each dentil in place. The technique they are imagining certainly would demand patience and a talent closer to madness than skill. They would be surprised to find that making the dentils took me 20 minutes.

To make the dentil moldings, I began with three lengths each of 1x3 and 1x4 stock (for the three lengths of the entablature). I have no respect for the edges you get on lumberyard stock, so it's my habit to joint such stock, rip it to width and joint the other side. I ripped the 1x3s to 2¼ in. and the 1x4s to 3 in., then glued the narrower stock (for the dentils) to the wider stock (the frieze board), aligning the top edges. Then I cut between the dentils with a dado blade in a radial-arm saw, using a gauge block screwed to the fence to keep the spacing accurate.

I primed and topcoated all moldings in the shop, brought them to the site and nailed them in place. The frieze is blocked out to allow sidewall shingles to be tucked underneath, and it runs over the corner boards. A bed molding under the dentil molding returned the fascia to the frieze below, and the entablature was complete.

Salvaged sidelights—The owner had salvaged the sidelights long ago from a two-hundred-year-old house. Little about their overall appearance suggests their age, but some anonymous artisan had sawed his stock and planed the parts in the quiet of a colonial New England village. He chiseled the joints and secured them with wood pegs, and two centuries later his sash was returned to use with a few repairs, new glazing and new paint. It was a particular pleasure to give new life to the work of an earlier century.

The sidelights were 8 in. shorter than the stock door, which caused a problem. To line up the heads, I built the sidelight frames with a filler panel below the sill. The sidelight frames and the stock door jambs are finished with beaded casings, or architraves. An apron runs the width of the entryway between the corner boards, and is raised a fraction above the landing to prevent moisture from being trapped underneath.

I made the fluting for the pilasters with a molding knife on the table saw. Each pilaster is fitted with a plinth block and a capital. The plinth is no more than a block with a standard base cap mitered around the top. The capital ends under the frieze, as would the top of a column supporting the entablature of a portico. Strict attention to form would require that the capitals be an independent part of each pilaster, but a more graceful appearance was achieved by running the upper capital molding across the tops of the sidelights and door (photo p. 21).

We used solid oak sills in the new entryway for both the outer door and the new inside door. Although they are more expensive, I prefer solid oak to laminated oak sills because laminated sills will eventually come apart.

Assessing the entryway—One question still concerned me, even when the entryway was finished: How will a formal Greek Revival entry look on a Cape Cod cottage? I had made detail drawings and perspectives a dozen times over, but nothing tells the truth better than a casual view from a distance. And coming down the long driveway in the last days of a New England summer, the entryway does just what the owner had in mind. It is neither obtrusive nor frivolous, but looks if it has belonged to New England for years. □

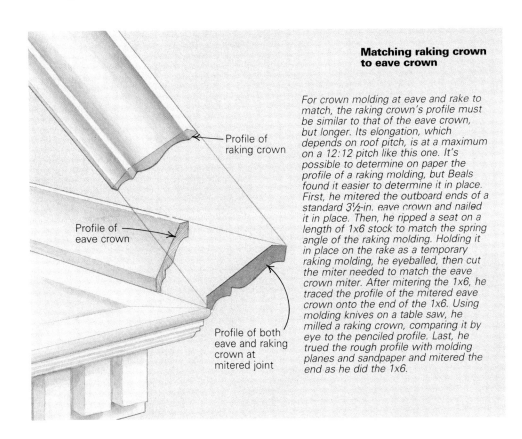

Matching raking crown to eave crown

For crown molding at eave and rake to match, the raking crown's profile must be similar to that of the eave crown, but longer. Its elongation, which depends on roof pitch, is at a maximum on a 12:12 pitch like this one. It's possible to determine on paper the profile of a raking molding, but Beals found it easier to determine it in place. First, he mitered the outboard ends of a standard 3½-in. eave crown and nailed it in place. Then, he ripped a seat on a length of 1x6 stock to match the spring angle of the raking molding. Holding it in place on the rake as a temporary raking molding, he eyeballed, then cut the miter needed to match the eave crown miter. After mitering the 1x6, he traced the profile of the mitered eave crown onto the end of the 1x6. Using molding knives on a table saw, he milled a raking crown, comparing it by eye to the penciled profile. Last, he trued the rough profile with molding planes and sandpaper and mitered the end as he did the 1x6.

Joseph Beals is a designer and builder who lives in Marshfield, Massachusetts.

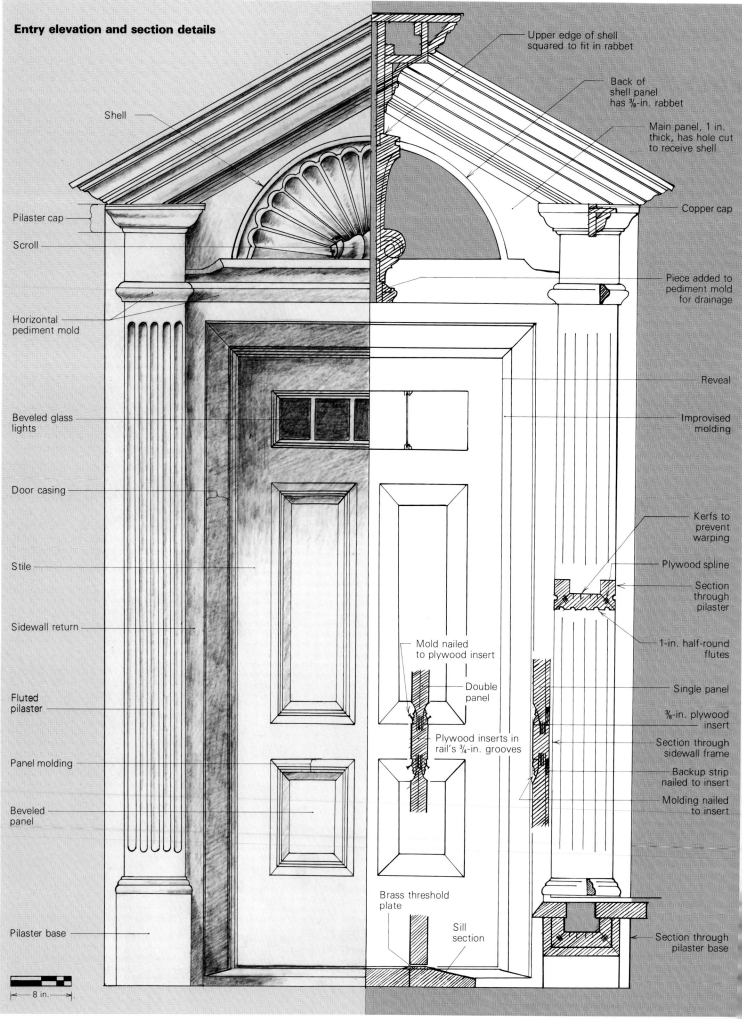

Formal Entryway

A challenging adaptation of traditional designs

by Sam Bush

It was my pleasure to have been working for Pat Dillon, of Canby, Ore., when Mr. and Mrs. Alfred Herman asked him to build a new entry for their home in Portland. The 18-year-old existing door was not particularly well designed or constructed, and the Hermans wanted to replace it with one that incorporated various traditional elements. Pat often said that the doorway is the most important part of a residence because it introduces the people who live there. Pat and I liked the Hermans, and wanted them to be introduced properly.

We were a good combination of clients and carpenters for work of this sort. Pat, of Canby, Ore., is an experienced contractor with high standards for everything he undertakes. My own background is primarily in furniture, cabinetmaking and woodcarving, but I enjoy challenging finish work. Pat had earned the confidence of the Hermans before I'd come along. It was encouraging to know that they trusted us to do a good job and that our workmanship would be appreciated.

Design and planning—Helen Herman had a pretty clear idea of how she wanted the entry to look: pedimented, with fluted pilasters on either side, and paneled sidewalls. The basic concept was simple elegance. So Pat and I did our research, visiting, photographing and sketching many doorways. This was great fun for me, as I had studied Colonial and Federal work all over New England. We presented our ideas, and after many conversations with Helen we arrived at the proportions and lines that captured what she had in mind. The final design is a personal one, though it is based on traditional elements. We consulted on every part, right down to the panel moldings, which were created with shaper cutters profiled to our drawings. Relying on the good eyes of three people, we strove for balance and attractiveness throughout.

There were two stages of construction: making all the parts in the shop (which took six weeks), and installing them (which took seven days). First came shop drawings and calculations. Pat had made discreet investigations of the substructure of the existing entry without disrupting its appearance too much. We used his measurements for planning and ordering lumber.

For everything but the shell, we used clear, vertical-grain old-stand fir, a wood readily available in the Northwest. It's expensive, but great to work with, except for a tendency to splinter. (One afternoon, as I was working to extract one of a long line of slivers, Pat jokingly accused me of trying to take home wood from the job.)

The frames—Our first job was making the door stiles and rails. All of it was made of 2¼-in. stock, and we chose the pieces very carefully to be sure they would stay straight after installation. All the joints were deep mortise and tenon, haunched top and bottom for strength, and glued. There was also a ¾-in. wide groove run up the inside of every framing member in the lower part of the

A blend of New England architectural elements, the new entryway features custom moldings, a shell carving and a copper-clad roof.

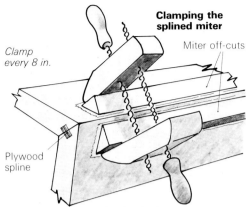

Clamping the splined miter

Clamp every 8 in.

Miter off-cuts

Plywood spline

Layer of newspaper glued to both miter off-cuts and surface allows for easy removal after clamps come off.

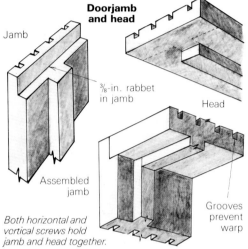

Doorjamb and head

Jamb

3/8-in. rabbet in jamb

Head

Assembled jamb

Grooves prevent warp

Both horizontal and vertical screws hold jamb and head together.

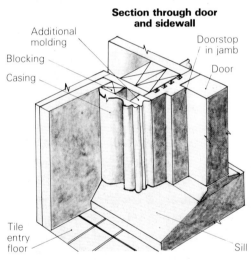

Section through door and sidewall

Additional molding

Blocking

Casing

Doorstop in jamb

Door

Tile entry floor

Sill

Door-frame construction is haunched mortise and tenon; grooves (photo below) will receive plywood inserts for attaching panel moldings. Long ears on door sill (photo bottom) will support door casing. Molding to receive lower right panels is already in place. Other joinery details are shown in the drawings at left.

correspond with the plywood inserts. With the other cutter, we ran the molding that would hold the panels in place without inhibiting their natural movement. It was fairly easy to miter the molding on the chop saw and install the panels, which had been treated with wood preservative and primer paint. (All the parts were treated on both sides with two coats of Houston's #3 preservative and one coat of Kelley-Moore exterior primer before final installation.) We used Titebond to glue the molding to the plywood inserts before setting two finish nails on the horizontals and three on the verticals.

The other custom molding for this job was the 4½-in. wide door casing, similar to the panel molding, but larger. We were able to make some of it with shaper and router cutters, but the main line was really handmade, with a few table-saw cuts and a lot of scraping and planing.

Jamb, cornice and pilasters—We used solid stock for the door jamb, and grooved it on the back to prevent possible warp. The sill was screwed to the rabbeted jamb, flat under the door and sloping 5° on the outside, with long ears left on either side for the casings to rest on (photo left). After the jamb and sill were glued and screwed, the door was dimensioned and beveled, leaving a 3/32-in. space all around to allow for paint. Hanging involved setting three brass-plated ball-bearing butts, 4½ in. by 4½ in., using a Stanley door guide and a router to make the hinge mortises. We also fitted the brass latch and the Baldwin Lexington-pattern lock hardware, and I can tell you it's an exacting task to get the precise mechanism just right and working smoothly. (And this is no time to drill a hole in the wrong place.) The striker plate was not set into the jamb until later. Making up and nailing on the molding that holds in the beveled glass window completed the door work.

Next, we drew a full-sized plan, on plywood, of one-half the triangular frieze and cornice area. We figured out the length of the roof and the sizes of the box structure to which the cornice would be attached, the main panel, the shell and its moldings. With this information we were able to make up all these parts and get them waterproofed and primed. Since the roof boards over the cornice and the pilaster caps were to be sheathed in copper, they were sent out to the sheet-metal shop for fitting.

The last step before installing the entryway was making the pilasters, which were mitered and splined like the sidewall returns. The major difficulty here was the fluting, which—like everything else—had to be just right: five 1-in. diameter half-round grooves on the face, with exactly ½ in. between them, and one groove on each side surface. I made these on the shaper, relying on a sample block to make the settings. Since the fluting didn't extend to the end of the pilasters, I had to drop the work onto the cutter and pull it off for each cut in accordance with layout lines on the work and the fence. I ran the cuts from both edges. The middle one was quite a reach for Pat's shaper, but everything worked out. The cutter left a peculiar-looking shape where it came out of the work, so I finished the ends of each flute with my carving gouges. After

door, to hold the plywood inserts to which panel moldings would be attached. We didn't put a groove in the top panel of the door, since a beveled-glass window would be located there.

Next came the framing for the ceiling and sidewalls, which was made from 1½-in stock, with mortise-and-tenon construction like the door. We ran 3/8-in. grooves for plywood inserts that would be arranged slightly differently from those in the door. These frames were assembled with a wider stile on the inside so that when the door casing was installed, overlapping it, the remainder showing would be the same size as the outside piece. We also built mitered returns on sidewall and ceiling frames where they would turn and lie on the face sheathing of the building. We decided early on to use as few nails as possible, knowing their tendency to move and show through, so we joined these miters with ¼-in. plywood splines and glued them with temporary clamp blocks made from the miter off-cuts, as shown in the drawing, top left.

Panels and molding—Door panels could have been made of single 2¼-in. pieces, but we chose to use two thicknesses back-to-back so that they could adjust individually to indoor and outdoor conditions, avoiding the tendency to warp. Each 1⅛-in. thickness was made of two edge-glued boards; we ran the edges on the shaper with a special glue-joint cutter to prevent creep along the joint. Almost every joint in the job was glued with Weldwood resorcinol waterproof glue.

We shaped the panels with a 3/8-in. tongue to

this we cut the pilasters to length and mounted a three-part molding at the top of each. We were finally done in the shop.

Installing the door—While we were doing all this, the tile men had been at the Hermans' laying a 4-in. by 8-in. quarry tile surface on the floor of the new entry. Our first task was to demolish the original finish work. The existing sidewalls were ½-in. plywood with molding nailed on to make them look paneled. Fitting all over was casual, and construction choices were ill considered. In fact, a great deal of water had gotten into the trim and rotted it.

Removing the old doorway was last, and meant we could start rebuilding. It also meant we would be there until the new door was in, since we weren't about to leave for the evening with a big hole in the house. Setting the jamb and sill assembly had to be particularly accurate to fit both the existing interior and the new exterior parts, but the installation was done conventionally. To help take the weight of this heavy door, we ran 3-in. #8 screws through the jamb into the rough framing. These were in the hinge mortises where they would be covered. With the door in place and swinging, I marked and mortised the jamb for the striker plate and dust box. I took special care here, wanting as a matter of pride to have the door close both easily and tightly.

Entry paneling and casing—After the door was installed, we started work on the ceiling unit, the panel of which was pierced to accommodate the overhead lamp. This frame had to be set an exact and uniform distance from the jamb so the door casing would fit evenly. We then leveled, shimmed and nailed it along its left and right edges with as few 16d galvanized finish nails as possible.

The sidewalls were next. Like the ceiling, they were located parallel to the jamb—no margin for error here. They were also scribed to fit both the ceiling (covering its nails) and the pitched tile floor, while showing as a plumb line on the outside face. To prevent rot, we held all our finish work ¼ in. off the tile.

It was at this point that we discovered the only serious mistake of the job. Somewhere back in the calculation stage, I had erred by an inch in the sidewall measurements, making them narrower than they should have been. While the casing fit to the walls, the visible vertical stiles were now of uneven widths, and the balanced effect was spoiled. They say the *real* skill of a carpenter shows in how he gets himself out of the mistakes he gets himself into. Pat earned his money that day, inventing a piece of molded stock that brought out the surface of the jamb and solved the problem. This disaster cost us time, materials and more than a few grey hairs, but as sometimes happens, the adjustment turned out to be an improvement. The additional molding looked natural and made the doorway look even better.

With sidewalls and improvised molding in, and corresponding blocking next to the wall, it was my job to set the handmade casing. There was a lot to keep in mind while I was working: There had to be an even reveal along the door and perfect fit to the wall; both of the miters had to be tight and meet the corners of the wall and ceiling frames; the side pieces had to be cut at 5° on the bottom to sit on the angled sill—and there was no extra stock.

Around the entry—Next, we moved to the face of the house. We determined where the exterior vertical sticks would rest, and cut the cedar siding to the chalk line using a portable circular saw against a straight board. Likewise we made the angled cuts for the peak, after carefully measuring to ensure that the peak was plumb above the ceiling midline. When the side pieces were laid, they were tight against the siding, with their surfaces in line with the sidewall returns. They were also bedded in a thick corner bead of butyl caulking, which sealed the joint. We later ran another tiny bead of colored caulking along the surface of the siding where it met the side pieces.

Next, we notched the roof boards to fit over the siding, and toenailed them into the sheathing. We screwed blocking under their outer edges, then fit them into their copper covers, working a long flange of the metal up behind the siding. We bent the covers around the roof front edge, where we discreetly nailed it into the blocking with four copper nails. An overlapping, interlocking joint between copper sheets along its apex rendered the roof watertight. Then we mounted the 1-in. thick main panel on shims that brought it out flush with the 1½-in. trim. The back of the shell carving would eventually fit into the space behind the panel, so the shims were placed accordingly.

With the background and roof in place, we fastened the pilasters plumb, leaving an equal width of vertical boards on both sides of each one. The top moldings were already attached, so we needed only to add the angled cap pieces with their matching copper covers. At the base we nailed on the 1-in. thick spline-mitered base assemblies, after first scribing them to fit the tile. We cut and installed the molding along the tops of the bases to complete the pilasters.

The cornice box went up next, against the roof and backwall blocking, mitered at the top. Putting up the detailed cornice moldings was just about the most trying part of the whole job. A combination of factors—difficult miters inside and outside, close reveals, scribing to the siding, no extra material—left no room for error. But patience and a sharp block plane prevailed, and the parts fit as they were meant to. We used needlepoints, 1¼-in. galvanized finish nails, to fasten these moldings. They hold tightly and don't split the thin molding edges. Driven flush with the surface, they hold paint and don't show. We then installed the two-part detail inside the cornice box and above the shell. The moldings were mitered at the top, and died out against the pilaster caps. The last of the molding work was the strong horizontal piece right above the entry. Compared to the others, this one was simple because it was level, with regular 45° miters. Both Pat and I were glad to be done up above. We had made an unbelievable number of trips up and down the ladders.

The installation was completed except for the shell. We agreed on the lines of its elliptical shape and put up a cardboard mockup to see how it looked, and to get Helen's approval. There had been talk of a fan light or other details above the door in the beginning, but it was the shell that really fit into Helen's idea of the entry. That made me feel pretty good, because it had been my suggestion.

The shell—Back at the shop we used resorcinol to glue up the block from three layers of clear ponderosa pine, which carves nicely. We paid close attention to getting truly flat surfaces, edge and face, laying the boards so their edges did not

To install the entry, the old siding was first removed, left. At right, the entryway complete but for the shell carving—the cardboard mock-up holds its place.

line up. The bottom piece was longer than the main area and was made with an angled top edge glued in an angled rabbet, so the ears on the edges of the shell would have a draining surface. Pat had the further good idea of adding another 1-in. piece to the scroll areas so it could project out beyond the rest. This we did, and it added greatly to the finished effect.

I did the carving at home on my basement bench in three-and-a-half days of work. The photos below illustrate the sequence of events. After carefully tracing the pattern onto the wood from the full-sized drawing with carbon paper, I drilled a series of holes of uniform depth very close to each other around the scroll area. This made it possible to remove the wood with chisel and mallet, working always from the outside to the middle and having the long chips break off against the scroll. When the drilled holes started to disappear, it was time to quit at the middle. Other than running my hand back and forth across the dish-shaped depression, feeling for bumps, there was no guide or template to measure the curve. The dark glue-lines showing through were a big help in this respect. The bottom of this area was left angled about 12° for drainage.

The scroll itself was next, a careful sculpting that involved much checking and rechecking to make sure the two halves came out the same size with natural-looking spirals. I established the ends first and then shaped the rest of the block into a balanced, compound-curving mass that looked a bit like a football with its ends cut off. On this bulk, I laid out the lines for the center bump and the ribbon-like strips at the ends. The careful modeling of the whole block first made it relatively easy to carve in the molded shape and have it come out right. The ends, however, were just plain work. They were difficult to visualize because the spiral shape climbs upward to the tip, and does not lie in a flat plane.

With the scroll complete, I laid out an even spacing around the inside and extended lines from the perimeter to the center with a flexible straightedge. Guided by these sets of tapering lines, I carved the shell flutes with a long-bend gouge. Here again the lamination lines were helpful in establishing depth. The ridges between the flutes were the only part of the dished shape that remained, and I carved these with a slightly concave surface. Then, using a router and straight mortising bit, I reduced the area between the scallops and the edge to a depth of $\frac{3}{8}$ in., trimming up the lines with carving tools.

The next move was bandsawing out the block's elliptical shape and the curving ears on each end. I did it with the table on a 12° angle to provide the needed pitch. At this time I also routed a 1-in. deep rim around the back, to allow weathertight mounting of the shell. I was then able to carve the deep concave molding of the base, and carefully sand the whole piece with papers ranging from 60 grit to 120 grit. After a thorough waterproofing during which the wood was lightly heated between coats to maximize absorption, the shell was primed.

Back at the Hermans, mounting involved cutting a hole to receive the shell in the background panel with a reciprocating saw. That allowed the shell's rim and base to lie on the surface of the background panel, so that its profile could be marked exactly. The shell overlapped the panel by 2 in. We routed out $\frac{3}{8}$ in. from the 2-in. overlap with a hand-held router, trimming to the marked outline with chisels where necessary. This left a wide, flat rabbet into which the shell block fit. The bottom edge of its base was worked to correspond with the angled surface of the molding on which it sits. We ran two wide beads of butyl caulk into the rabbet, and permanently mounted the shell with ten large galvanized finish nails in predrilled holes. A thin bead of caulking along the joint between the shell and background panel completed the job.

While I was finishing the shell, Pat got the weatherstrippers to the job to install the brass threshold and brass interlocking trim all around the door. He also installed a copper mail chute the sheet-metal people had made, to carry mail from an extruded brass mail slot to a box located in a closet. This box collects the mail, and is fitted with a door so there isn't heat loss through the mail slot.

That completed our work, except for Pat's overseeing the subcontractors. The concrete men formed up and poured a new set of steps up the bank, and gave them an attractive exposed-aggregate surface. Then the wrought-iron people installed the new railings. And the painters came and finished our work with two coats of exterior oil-base white. For us it was a very successful and satisfying conclusion to a big job. The Hermans were certainly pleased, and said so, and that was our bonus. □

Sam Bush heads the wood program at the Oregon School of Arts and Crafts in Portland.

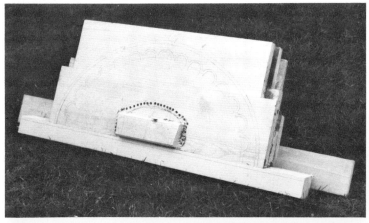
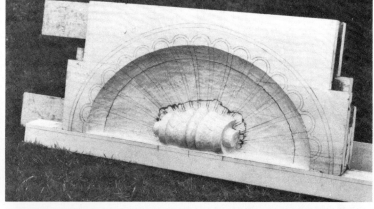

Making the shell begins with gluing up three layers of ponderosa pine, edges staggered for strength, top left. Small block at front is for scroll; holes will aid in carving flutes. Top right, fluting is laid out in dished-out shells; scroll carving is complete. With flutes carved, above left, semicircular groove is routed and block is bandsawn to shape, above right. A concave molding base completes the job.

French-Door Retrofit
Opening up a wall to bring the outdoors in

by David Strawderman

What we know today as French doors were originally considered tall casement windows that simply reached to the floor. They appeared in the late 17th century at Versailles—Louis XIV's grandiose headquarters southwest of Paris. There they overlooked immense gardens that required the rerouting of a river for adequate irrigation.

French doors are still a gracious architectural element that can enhance the appreciation of a garden—no matter what its size. A pair of them make the connection more immediate than even a large window. My clients for the job illustrated in this article, Luis and Carol Fondavila, wanted a closer link between their breakfast/dining area and their beautifully landscaped backyard. Here's how I removed the existing double-hung window, created a 6-ft. wide opening and filled it with a pair of new French doors.

Headers and rough openings—Creating a new opening in a bearing wall requires that you shift the load from the existing studs to a post-and-beam carrier. The beam, or header, is typically a 4x timber or a couple of 2xs with a piece of ½-in. plywood sandwiched between. Then the entire assembly is spiked together with 16d nails. The depth of the header depends upon the span and load, and the common rule of thumb for single-story dwellings is that 1 in. of header depth will span 1 ft. For example, an 8-in. deep header will span 8 ft. A house with a typical 8-ft. ceiling and standard 6-ft. 8-in. doors will usually have enough space between the top plates and the door framing for a 12-in. deep header.

But what about two-story houses, like this one? I use common sense, look at what's in the wall and act accordingly. In this case, the existing 4-ft. wide window had a single 2x8 header. I decided that a double 2x8 would be adequate to span 6 ft. If you have any doubts about this kind of a calculation, however, have an engineer or architect size the header.

Another structural consideration besides the header depth is the foundation bearing capac-

French doors. Traditional French doors consist of a pair of multi-light doors, both of which are operable. The primary door (in this pair, the one on the left) carries the lockset. The secondary door holds the strike plate, and is held fast by deadbolts.

ity. The loads from the new header are passed by way of the trimmer studs to the foundation. This creates "point loads" on the foundation, and if they're considerable they can crack an otherwise adequate footing or a foundation atop weak soils. The Fondavilas' house has a massive foundation with 10-in. stemwalls in perfect condition. I was satisfied that the new loads from a 6-ft. long header weren't going to cause a problem. If the header span is longer than about 7 ft., however, I consult an engineer or an architect if I'm at all in doubt about the integrity of the structure. It's important that this be addressed *before* any holes are cut in the walls. Reinforcing the foundation at the post-bearing points can be messy, expensive and time-consuming, especially after the loads have already been altered.

Once I've considered the structural part of the equation, I turn my attention to utility obstructions. Telephone and television lines are minor items that are easily dealt with. Electrical wires, gas and water lines hidden in the wall can present more serious problems. First I check the exterior wall for utility entrances. If a gas line or an electric service is entering the house at precisely the spot where my client wants a new set of doors, I explain that the utility companies will have to relocate them, thus adding considerable expense. Perhaps the doors can be moved a little to accommodate the obstructions.

The interior wall offers clues to probable electrical wiring paths. Outlets typically have to be relocated, and light switches moved to the side of the new opening. Sometimes a look above the proposed door location, in the attic for example, will reveal evidence of wiring in the stud bays that will be affected. I also check in the basement or crawl space.

The most difficult utilities to relocate are gas, water and waste lines. Fortunately, most plumbing lines run through interior walls. If you are putting doors in a wall that already has a window, chances are you won't encounter plumbing lines in the portion of wall below the window. There are, of course, exceptions to this, such as long trap arms from a sink, or vent pipes that take a horizontal jog across a wall. Implanting a set of doors in a blank wall runs a greater risk of running into water and gas lines. To assess the likelihood of finding pipes, take a look in the crawl space or basement under the section of wall in question. Again, you may want to move a door a few feet this way or that to avoid relocating pipes. Upstairs bathrooms and vents in the roof can give you additional clues as you learn about where the pipes may be buried in the walls. Keep in mind that the more obstructions there are, the higher the cost will be. When I am satisfied that cutting and framing the rough opening is feasible, I address the design of the finished unit.

Style and costs—A pair of French doors engage each other in three basic ways: the hinged pair; a hinged door with a fixed sidelight panel; and a sliding pair. Deciding which kind to install is usually a matter of style.

For the classic look and a generous doorway that can be opened to the breezes, nothing compares with the traditional pair of hinged, true divided-light doors (photo previous page). Their delicate muntin grids are compatible with the windows of most homes built before the '50s. On the downside, a pair of hinged doors are tough to install and they are often considered the Achilles' heel of weatherization. It can be hard to keep wind-driven rain from getting through the gap between the doors, though properly installed compression weatherstripping can do the job. Another way to protect them from the weather is to install an awning-type canopy over them, such as the one designed by Bill Mastin (*FHB* #63, p. 56).

French doors with fixed sidelights are easier to weatherize, and I charge considerably less to install them because they have only one operable door. They can be a good solution when space is tight.

Sliding French doors can be very wide because they ride on rollers. Sliders aren't susceptible to being blown shut by the wind and they don't get in the way when they're open. Sliders without muntins are compatible with modern architectural styles.

The main floor of the Fondavilas' house, built in the '20s, already had several pairs of 7-ft. tall French doors. Even though they were no longer available off-the-shelf, my clients wished to duplicate them, as well as the hardware.

When I don't have to match existing doors, I buy standard units from my local supplier. There I can select from stock doors that have from 10 to 15 lights per door, and are either 2 ft. 6 in. or 3 ft. wide by 6 ft. 8 in. tall. These doors cost between $140 and $200 each. They are made of solid vertical-grain Douglas fir, and their single-glazed, tempered lights are double-bedded, which means they are glazed on both sides to help keep out the rain. I can buy the same doors prehung with jambs and a threshold for around $800 from local door shops, or pay about double that for some nationally known brands. For about $120 worth of material I can make my own jamb set and prehang the doors in less than half a day.

Doors that have to be made from scratch cost a lot more. One bid for the Fondavila doors came in at $650 apiece. I settled on a pair that cost $375 per door. Prehung custom door units rise in cost accordingly. In any case, locksets are extra.

If you order a custom set of prehung doors, your supplier will need to know the details of the doors, the vertical dimension from the bottom of the threshold to the top of the jamb, the horizontal dimension from the outside edges of the jambs, and the depth of the jamb from the finished interior wall to the finished exterior. The rough opening should allow a ¼-in. gap at each side and at the top.

Hardware—A pair of French doors may open out or in, and the door you open first is called the primary door (or the active leaf). The primary door holds the entry hardware, while the secondary door is secured to the upper jamb and threshold with sliding bolts. These bolts can be surface-mounted on the interior door side, such as those in the Fondavilas' house, or mortised into the door edge. The entry and deadbolt sets are the same as those used for single doors, and their strike plates are secured to the secondary door. The secondary door typically has nonoperable knobs to match the active leaf.

The closed doors need a stop where they meet in the middle. There are two basic solutions and the T-astragal is by far the most typical (top drawing, p. 32). This molding strip is secured to one of the door edges. It's best to orient the crossbar of the T to the exterior

Out with the old. A reciprocating saw makes short work of old plaster, studs and the nails that hold them together.

Using a Carborundum masonry blade ensures a straight, clean cut in a stucco wall.

The new header is inserted into its cavity in the old wall. Note how the cripple studs over the window have been cut in a horizontal line so that they can bear equally on the header.

New trimmer studs support the weight of the header and its load, and define the edge of the rough opening. Here the door jambs are lifted into position and aligned with the interior wall.

side so that it will conceal the latchbolt and protect against the weather. The astragal may be on the secondary or primary door, depending on whether it swings in or out.

The alternative is a set of doors with interlocking, rabbeted edges. Although this configuration is elegant, the entry hardware for it is limited. The only company I know of that makes locksets and strikeplates suitable for rabbeted doors is Baldwin Hardware Corp. (P. O. Box 15048, Reading, Pa. 19612; 215-777-7811).

Jamb assembly—The jambs I use are made out of fir, and they are typically 1⅛ in. thick with a ½-in. deep rabbet along one edge to create an integral door stop. The interior edge of most jambs is flush with the interior finished wall, and the exterior edge is flush with the exterior wall. The width normally falls between 4¼ in. and 5¾ in. With the Fondavila job I had a 9½-in. deep wall, so I used 5¾-in. wide jambs and made up the difference with trim and stucco mold (drawing, p. 33).

A flat, open area is useful for layout and building the jamb set. I begin with the head jamb by marking off the dimensions of the doors, the astragal and its space and the hinge spaces (top drawing, p. 32) on a piece of jamb stock. If I'm installing a threshold, its length is equal to the outside dimension of the assembled jamb plus the length of the ears. The side jamb fits into a tapered rabbet at the ends of the threshold, next to the ears (bottom drawing, p. 32).

The length of the side jamb equals the door height, plus the thickness of the top jamb, the depth of the tapered rabbet in the threshold and a ⅛-in. gap above and below the door for clearance. If I'm installing a sweep or a gasket at the bottom of the door, I'll adjust that gap accordingly.

To join the jambs at the top corners, I cut a rabbet in the end of the side jambs and screw them to each end of the head jamb with three 3-in. drywall screws (bottom drawing). Before assembly, I lay down a bead of Polyseamseal caulk in the joint. The caulk serves as both a glue and a waterproofing agent (Polyseamseal, Darworth Co., 50 Tower La., Avon, Conn. 06001; 800-624-7767).

Incidentally, before I commit myself to a certain door height, I make sure the doors are square. They aren't always, and finding out after cutting the jambs is no fun.

Once I've assembled the jamb frame, I square it up and reinforce it with some diagonal 1x2s screwed into the edges of the jambs opposite the hinge side. The jamb frame can now be moved around pretty easily, and I lift it up onto the bench for easier access.

I typically install compression weatherstripping along the inside edge of the doorstops. I need to do this before the hinge gains are routed to allow space for the thickness of the weatherstripping. To cut the grooves for the weatherstripping, I use a slick little router made expressly for the purpose by Weatherbead Insualtion Systems Inc. (5321 Derry Ave. F, Agoura Hills, Calif. 91301; 800-966-0159). For more on this tool, refer to my review in *FHB* #60, p. 92.

While the jamb is on the bench, I bring the doors alongside so I can mortise the hinges. I use a Bosch 83038 Router Template (Robert Bosch Power Tool Corp., One Hundred Bosch Blvd., New Bern, N. C. 28562-4097; 919-636-4200) for this operation. With it, I can rout the gains for both the jambs and the doors at the same time. I use three hinges per door.

Next I affix the T-astragal to the correct door, and put a 3° bevel on the edge of the door that meets it. Because small misalignments are difficult to correct after installation, I wait until the unit is installed to position the lockset and security bolts.

If the client has decided that the doors should be painted instead of varnished or stained, now is the time to prepare them for priming by removing all putty and caulk around the glass. After sanding the doors and jambs with 120-grit sandpaper, I apply two coats of Kilz Oil Base Primer (Masterchem Industries Inc., P. O. Box 368, Barnhart, Mo. 63012; 800-325-3552).

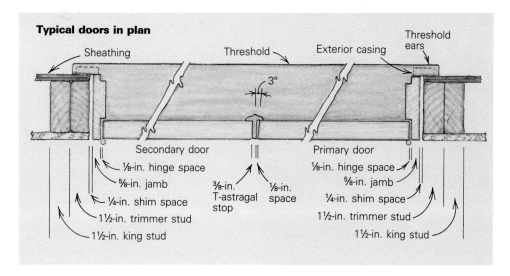

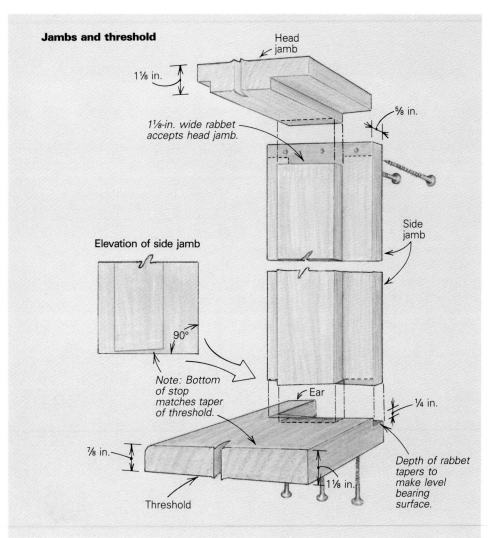

Opening up the wall—I begin work on the wall from the inside. This allows me to reposition any utilities, remove wallboard or plaster, and sometimes even assemble the new framing before breaching the exterior wall.

I locate the center of the new opening, and lay out the width of the door jambs, adding ¼ in. on each side. For an 8-ft. wall I draw vertical lines floor to ceiling on both sides. Then I use a reciprocating saw with a short plaster-cutting blade to cut the drywall or plaster. I hold the saw at a shallow angle to avoid hidden electrical wires (use a dust mask and goggles). If the vibration from the saw is cracking the plaster, I screw a 1x3 to the wall outside the cut line to hold the plaster together during the cut.

On an 8-ft. wall I remove the drywall or plaster all the way to the ceiling, and using a utility knife, score the inside corner where the wall meets the ceiling. At the Fondavilas' house, however, I stopped the cut at the point where the top of the new header would meet the old studs to avoid having to make a much larger patch in the 10-ft. wall (photo, p. 30). The wall's innards are now exposed, and if pipes or wires need to be moved, now is the time to do so.

Here's the typical sequence I follow when I'm making a rough opening 7 ft. wide or less. After removing the drywall or plaster, I use my reciprocating saw to cut in half the studs that need to be removed, and unless I find some evidence of concentrated loading from above (like a pinched sawblade as I crosscut a stud), I don't bother with shoring up the ceiling. A metal-cutting blade in the reciprocating saw makes it easy to cut the siding and sheathing nails away from the studs as I remove them.

Next I install the king studs on either side of the opening, slipping them into the wall cavity. I run a bead of Polyseamseal on the exterior edges of the new studs, and nail through the existing wall material into them after toe-nailing the studs to the top and bottom plates. The distance between the king studs should be 3½ in. more than the width of the finished unit. This distance equals the length of the new header.

In an upper corner, I make a notch in the drywall or plaster to allow easy insertion of the header. I tack the trimmer stud opposite the notch to its king stud, leaning its bottom out to allow ample clearance for the header. Then I hoist the header into place, put the trimmer under the other end of the header and nail it off. I tap the slanted trimmer stud into its final position and nail it securely to the king stud. If the new header doesn't reach all the way to the top plate, I put in blocks or cripple studs to restore the load path of the old studs. All the new and old framing members should be toenailed securely to one another.

If the opening is over 7 ft., I put up some temporary shoring a couple of feet in from the wall to help carry the weight of any ceiling joists that might be affected by the removal of the old studs. The shoring consists of a temporary 2x6 top and bottom plate the length of the new opening, along with some studs wedged in place and tacked to stay put while the new header is installed (for more on retrofitting headers, see FHB #62, pp. 85-87).

That's the theory, and often it follows that order. The Fondavila job had its exceptions. The height of the walls made it impossible to install new king studs without tearing out a lot more wall, so I made the header longer than it would normally be and turned the existing studs on either side of the opening into the king studs. I made a built-up header out of a pair 2x8s with 2x8 blocks between them to help fill the stud cavity and lifted the header into position (top right photo, p. 31). The new header is toenailed to the old studs, and it bears on the new trimmer studs that frame the rough opening (bottom photo, p. 31).

Punching through—I begin the final process early in the day to ensure ample time to have a locked set of doors in place by evening. I begin by locating the corners of the opening on the outside of the house. A long ½-in. bit works nicely (I used a masonry bit for this job because of the stucco siding). Holes drilled, I move outside and snap chalklines for the cuts. They should be flush with the inside edge of the trimmer studs and the bottom edge of the header.

My helper, Larry Furniss, used a circular saw with a carborundum masonry blade to cut through the stucco finish (top left photo, p. 31). It's dirty work but cutting stucco this way ensures accuracy and leaves the remaining stucco undamaged. Then I used a reciprocating saw to sever the wood sheathing behind the stucco and any nails or studs that held this portion of the wall to the house.

Other exterior wall surfaces and molding details are cut differently. Cuts in horizontal

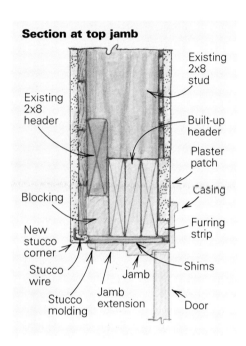

Section at top jamb

siding, for example, sometimes need to be a few inches outside the rough opening to accommodate recessed molding. Bullnosed stucco-returns with no exterior wood molding need to be cut flush, then taken back several inches so new stucco wire can be tied in.

Installing the jambs—After removing the doors, Larry and I moved the jamb frame into place and lifted it into the opening from the outside until its interior edge was flush with the plane of the wall (bottom photo, p. 31). I didn't put a threshold on the Fondavilas doors because they plan to have a mason create a cast-concrete sill and stoop to match the other doors. If the threshold is affixed to the jambs, however, its ears should be flush with the exterior finish. Check the jamb against the interior and exterior surfaces. Most walls have variations, so you'll have to average them out. When remodeling, I've found it better to conform to existing conditions than to follow my impulse and make everything plumb and level.

I wedge the jamb at the top to hold it in place without any nails while I hang the doors. Instead of leveling and squaring the frame, I use the doors as a guide to "squareness." If properly prehung, small in-and-out and up-and-down adjustments to the jambs will bring the doors into proper alignment (photo below). I shim the frame on each side 4 in. from the top and bottom, above and below the top and bottom hinges and near the center hinge. I shim the top jamb and threshold near each corner and at 2-ft. intervals. I used to secure the jambs with 16d coated finish nails, but have switched to countersunk 3-in. drywall screws. Once I've got the jambs anchored to the trimmers so that the doors operate properly, I install the lockset, deadbolt and surface bolts.

Touching up—I usually patch any holes that I've had to make in the interior walls with pieces of drywall. Because the Fondavilas have plaster-on-lath walls, however, I had my subcontractor use expanded metal lath affixed to the new header in order to anchor a plaster patch. Like many a plaster wall, these are kind of wavy, so I installed the new door casings and baseboards before the final coat of plaster. That allowed my plasterer to bring the finish coat right to the edge of the casings, filling up the gaps caused by the wavy walls.

On the outside, I sealed around the framing and jambs with foam insulation. Then I nailed on the exterior trim and stucco molding, and readied it for the stucco man with a couple of coats of primer.□

David Strawderman is a carpenter in Los Angeles, California. Photos by Charles Miller.

The jamb and doors are brought into square with one another by inserting shims and blocks between the trimmers, header and the jamb.

Trimming the Front Door
Ripped, mitered and reassembled interior moldings add a custom look to a front entry

by Richard Taub

Inspired entry. Nearly all the trim—pilasters, mullions, corner blocks and crown—comes from leftover interior poplar casing. The field to which the trim is applied and the cap on top are clear-pine stock.

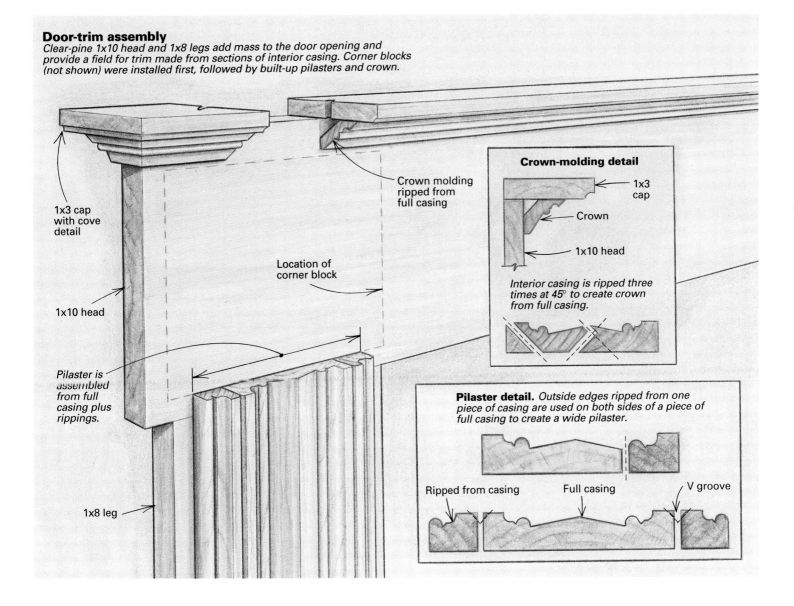

Door-trim assembly. Clear-pine 1x10 head and 1x8 legs add mass to the door opening and provide a field for trim made from sections of interior casing. Corner blocks (not shown) were installed first, followed by built-up pilasters and crown.

Crown-molding detail. Interior casing is ripped three times at 45° to create crown from full casing.

Pilaster detail. Outside edges ripped from one piece of casing are used on both sides of a piece of full casing to create a wide pilaster.

Consider the front door, if you will. There is something about this entrance that, if done tastefully, really invites a person into the home, eager to see more. Why, then, are so many front doors stamped in standard, prefabricated trim? You know what I mean: the same old dentils, sunbursts and broken pediments. They're usually southern yellow pine and can cost anywhere from $200 to $350. My partner, Steve Vichinsky, and I recently built a Victorian house with a large hip roof, an octagonal turret and a wraparound front porch. Considering all the work we had put into this house, I wanted to do something special with the front door.

My goal was to build something quickly that was pleasing to the eye and well made. I took a spin around town, looking at the ornate, older homes here in Amherst, Massachusetts, to get design ideas. A day and a half later I had a front door with custom trim (photo facing page) that looked great and was less expensive than the prefab stuff. Plus it was a lot of fun to make.

Interior trim used outside—A local millwork company had milled up all my interior trim for the windows and the doors (photo right). The casing was ¾-in. by 3¾-in. poplar. The first ⅜ in. on both sides of the face was milled square to the edges (detail drawing above). Next to these flat sections were ³⁄₁₆-in. coves and then a ⅛-in. V-groove. Two ⁵⁄₁₆-in. beads were next to the V-grooves. The rest of the trim came to a beveled peak in the center of the casing. I also had plinth blocks made for the doors, bullnose pieces to separate the casing from the plinth blocks, rosettes for above the doors and the windows, and detailed baseboard.

As I stood looking at a finished, trimmed-out room, it suddenly hit me: "Why can't I use the interior trim outside?" I stepped outside. Thinking of the houses I had seen, I knew this front door had to be built out in both width and depth to match the scale of the house. The interior trim alone wasn't beefy enough.

Fattening the trim—To give the opening more width, I nailed 1x8 clear-pine stock to the sides of the door and 1x10 over the top (drawing above). The head, or top piece, extends 1 in. past each leg. This gives a nice effect where the legs

Interior detail. Custom-milled corner blocks and casing inside the house inspired the design for the front-door exterior trim.

Cutting the corner-block components. Two 45° cuts make a triangle, and four of these triangles make a square corner block. The full width of the casing was used for the end corner blocks.

Another use for the biscuit joiner. After carefully cutting and aligning the four triangles, biscuits are inserted in slots to keep the joints tight through all of New England's weather cycles.

A closer look. Both the large corner blocks and the small diamonds were assembled with biscuits and epoxy (above), then nailed to the pine. Note how the overhanging head aligns with the siding (left).

meet the head, later accented by the siding. An overhanging head is a common feature of most fancy, exterior door trim. I now had a blank field of pine with which to work.

I tacked a short length of the 3¾-in. interior casing to a vertical 1x8. It looked good but was too narrow. I took another piece of casing and ripped on the table saw a 1-in. strip from each side. I added these strips, which included the ⅜-in. flat, the 3⁄16-in. cove and the 5⁄16-in. bead, to both sides of the first piece of casing. With a 1-in. strip on each side, the flat section facing in, the resulting ¾-in. flats felt heavy. So I set my table-saw blade at 45° and put a light chamfer on the edges of the rips and the casing. When I put the three pieces together again, the resulting V-grooves broke up the ¾-in. flats, and the width of the three pieces together looked good. All I needed was a full-length version for each side of the door.

Mitered corner blocks—Now I had to come up with something for the head. I thought of using the interior rosette corner blocks, but they were too small. Then I realized that I could make my own corner blocks, once again using the interior trim. I did some tests by putting together mitered sections of the casing with hot-melt glue (see sidebar facing page). Cutting the mitered pieces really isn't too tricky, but if you don't cut the pieces accurately, none of the profile lines will meet. With a sharp blade on my chopsaw, I cut the the casing across the face at 45° and again at the opposite 45° to make a right-triangle offcut (top photo, left). Four of these triangles together made a corner block. I tried different sizes and settled on the largest one, which had triangles cut from the full width of the casing. Large corner blocks seemed more dramatic and better suited to the house. I particularly liked the border that the ⅜-in. flat created around the perimeter of the corner block.

Now that I had my corner blocks, the trick was to join the four pieces permanently. Having just finished joining all the interior trim with my biscuit joiner, I realized that joining the mitered corner blocks with biscuits would be a perfect application for this tool (middle photo, left). I numbered the triangles and put them together face-down with the joints aligned. I marked one line across the center of each joint, cut the biscuit slots and then glued the corner block together (bottom right photo, left) with highly weather-resistant epoxy.

Laying out the field—Twelve hours later, when the epoxy had dried, I centered the two corner blocks over the 1x8 legs and nailed them in place. Then I measured the height of each corner block and cut two lengths of casing and four lengths of strips to fit tightly beneath the corner blocks. I centered and nailed a piece of casing on each 1x8 leg. Then I toenailed the smaller rippings on each side of the casing to draw them in tight and countersunk all the nails. Now I had two 5¾-in. pilasters on top of my 1x8 stock, leaving a ¾-in. reveal on each side. This gave me the depth that I wanted. The depth showed up more when the trim was painted two different colors (for more about color schemes, see *FHB* #74, pp.

Using hot-melt glue

When I was in the Wendell Castle Workshop (a furniture-making school outside Rochester, New York), I found, like most other woodworkers, that there is nothing like mocking up a piece, either full scale or scaled down, that you've worked out on paper. A mock-up quickly tells you whether or not the piece works technically and aesthetically.

Mock-ups and miter clamps—For mock-ups, hot-melt glue is incomparable. Once that gun gets hot, the glue melts out and sets quickly (60 seconds to 90 seconds). You squeeze out a bead of glue, stick your pieces together and move right along. Before long, you have a working model in front of you. The only problem with hot-melt glue is that it's so easy it can make you lazy. In situations where you need to see if more intricate joint details will work, such as a mortise-and-tenon joint instead of a bridle joint, put aside the glue gun and do the joint the way it should be done.

When I glue up long mitered joints, joining the sides of a 12-in. box, for example, I make triangular clamping blocks. I lay a bead or two of hot-melt glue on the blocks and stick them on each side of all the miter joints, centered from top to bottom (bottom photo, right). A C-clamp pulls the joint together as it grabs the clamping blocks. This puts clamping pressure directly on the center of the miter joint. It's a good idea to keep the glue gun hot, in case a block comes loose during a glue up. This could be a disaster, but if you're fast, you can remelt the glue, squeeze a bit more on, stick the block back down, let it cure for one minute and continue with the clamp up. The glue is so elastic that it easily pares off with a sharp chisel and can be further cleaned by some sanding.

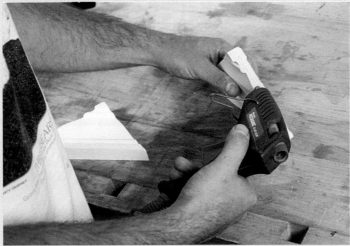

Melt, aim, shoot. This trigger-fed gun squeezes a hot bead of glue from its tip. The glue sets quickly as it cools, making it great for sticking together a prototype to see how it'll look.

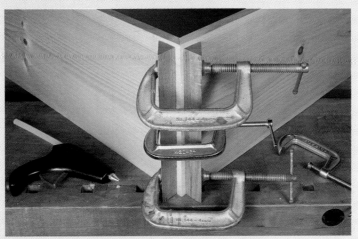

Temporary clamping blocks. To clamp miter joints, triangular clamping blocks can be made with hot-melt glue. The blocks can be easily removed later without damaging the mitered pieces.

If you plan to use clamp blocks to pull a piece together, and the piece needs a lot of clamping pressure, blocks glued on with hot-melt glue will never do. Once you start to really torque down on the clamps, the glue bond will break.

I always clamp up the pieces dry before using glue. Dry clamping tells me if the piece will pull together with minimal pressure or if it needs more serious persuasion. Also, I always have another plan ready if the glue fails to hold up under the clamp pressure—maybe some other conventional clamping means, such as band clamps (which happen to work well on mitered frame pieces of any dimension).

Glues and guns—Even though hot-melt glue is fairly strong, it cannot take the place of other wood glues. I don't recommend hot melt for gluing a finished piece because it cures so fast that you can't get a tight joint. Hot-melt glue just doesn't have the holding strength of yellow glue or epoxy.

As for glue guns, there are many on the market, and they are all fairly comparable in both design and price (about $10 to $25). Some, like my Master Mechanic 208-MM (top photo, left), are trigger-fed. This model loads easily; just slip the stick of glue in the back, and the trigger feeds it through the heating element. My gun gets extremely hot—about 380° F at the nozzle tip—so the glue squeezes out very nicely. I've found that the less-expensive models (usually not trigger-fed) don't melt the glue as nicely and won't allow those few extra seconds to work with the piece.

Hot-melt glue itself comes in an array of types and colors. The yellow sticks are basic, all-purpose hot-melt glue and are good for most household projects and repairs. I use the clear hot-melt glue, which is made for craft and hobby use, as well as for building or repairing stuff that doesn't take a lot of pressure or abuse. I wouldn't use it to repair a dining-room chair, but I would use it to fix a loose piece of trim. Hot-melt glue will bond to most porous materials within 90 seconds. There is also a brown hot-melt glue used as a fast-setting waterproof caulk for filling seams. I do not use this glue. Waterproof siliconized acrylic caulk gives you a lot of working time to fill a seam, and it cleans up easily. To get the same results with brown hot-melt glue, you must squeeze out the glue quickly and subject yourself to the heat of the glue as you press it into the seam with your finger. In other words, you would never get a good, clean, painless job. However, the brown hot-melt glue is useful for gluing something that might be exposed to moisture. —*R. T.*

40-45). The mullions between the sidelights and the door are much narrower, so I used the center beveled section of the casing ripped to 1½ in.

I now had an empty field of 1x10 trim above the door and the sidelights. Because the house has many angles and shapes, I thought diamonds might look nice in this empty field. So once again I started playing with mitered sections of the casing and came up with a smaller corner block. I made three identical corner blocks biscuited and glued together just like the large corner blocks. Turning them on their pointed ends gave me the diamond effect I was looking for. I nailed them onto the 1x10 head trim (bottom left photo, facing page), one above the center of the door and one over each mullion. All nails were countersunk to minimize stains; the painters would later fill these holes before painting.

Can you top this?—Now I needed to cap the head jamb. I routed a cove detail on a piece of 1x3 clear pine and nailed it, cove-side down, to the head. It made a little shelf. Then on the table saw I ripped another section of casing, making three passes with the blade at 45° to create a nice crown-molding detail (drawings, p. 35). I cut 45° compound miters on each end, centered this piece beneath the 1x3 and returned the crown into the sheathing to complete the trim. □

Richard Taub is a home-builder and furniture-maker in Amherst, Mass. Photos by Rich Ziegner except where noted.

Making an Insulated Door

This handsome entry keeps heat in while it keeps breezes out

by Irwin L. Post

Well-insulated, tightly fitting doors reduce fuel bills and make a house more comfortable. I recently built one for a client in Weston, Vt., who wanted a weathertight door that would complement her post-and-beam house. Her interior doors are made from knotty eastern white pine tongue-and-groove paneling and have custom-forged strap hinges. We decided that the inside of the new door should have the same appearance. For the exterior, we settled on five vertical boards surrounded by a border, all of knotty pine (photo right). The hinges, latch and knocker were made by C. Leigh Morrell at the West Village Forge in West Brattleboro, Vt. For a door this thick, you'll either have to modify standard locksets or special-order commercial models. I silver-soldered an extension to the rod that turns the deadbolt.

The heart of the door is a core of pine spacers and 1-in. thick beadboard polystyrene foam (R-value: 3.3), sandwiched between two pieces of ½-in. plywood. The plywood facing stiffens the core and makes it dimensionally stable and resistant to warping. One advantage of using a structural plywood core is that it allows great freedom in designing the finished surfaces on both sides of the door. You can glue stock of any shape or thickness to the plywood skin, creating, for example, freeform or geometric patterns. The facing boards on both sides of the door are splined together. As the drawing shows, the exterior face is made from nominal 1-in. paneling, bordered with 1x stiles and rails. I applied ¾-in. trim strips to the vertical edges, and these are overlapped by the facing materials on both sides. I chose to make the core height equal to the full height of the door and not to cover the top and bottom edges of the core with trim strips, since they are not visible.

Construction sequence—First, determine the size of the door. I made mine ¼ in. shorter and 3/16 in. narrower than the finished opening. Unless the door fits the opening precisely, air infiltration will render any amount of core insulation useless. Check to see that the opening is square. You can do this by measuring the diagonals—if they are the same length, then the opening is square. If it's not, you can either make the door fit the irregular opening or rework the jambs and casing to square things up.

Next, decide on the design for the door's inside and outside faces. It helps to make scale drawings to check the overall proportions of the various elements. Knowing the finished dimensions of the door, use your drawings to calculate the dimensions of the beadboard core.

Now, cut the two pieces of ½-in. plywood to the same dimensions as the core (the full height of the opening less ¼ in., the width of the opening less 1 11/16 in.). I used underlayment grade A-D ply with exterior glue. Cut 1x3 pine for spacers as shown in the drawing, step 1. For 1-in. beadboard, the spacers need to be a full 1 in. thick. Edge spacers can be made from short pieces if necessary, but they should be mitered and tightly butted to prevent air leaks. The diagonal spacers keep the door from twisting. The small blocks along the long sides are for extra reinforcement around the lockset. I glued a block on each side so I didn't have to keep track of which side would have the lockset and handle.

Glue the spacers to one piece of plywood. I used yellow glue. Be sure your worktable is absolutely flat, because a door built on a warped surface will be a warped or twisted door. A few 1¼-in. brads will keep the spacers from shifting while you apply clamps.

With a sharp knife or a razor, cut the beadboard to fit tightly between the spacers. Be sure not to leave any gaps. Now glue the second piece of plywood onto the spacers to complete the core (step 2). Clamps on the edges and cauls (cambered strips of wood which, when clamped at their edges, will exert downward pressure along their lengths) across the width of the door are a good way to get a bond. Use brads to keep the plywood from shifting as it's being clamped.

Attach the trim strips to the vertical edges (step 3). On this door, the strips were ¾-in. thick pine about 2¼ in. wide. I glued them on one at a time, and used a 2x4 to distribute the clamping pressure, as shown in the drawing. After both edges are glued on, trim them flush with the plywood on both sides. I used a ball-bearing pilot flush-cutting bit in my router, but a hand plane would work just as well.

While the glue is setting, prepare the paneling for the faces. I used splines and grooves to join adjacent knotty pine boards (step 4). You can use a dado head in a table saw to plow the grooves, or a slotting cutter in your router. The splines were cut on the table saw. The V-grooves that show on the outside of the door were made by beveling the boards with a hand plane.

Glue the paneling to the plywood core. I used no nails or screws because I wanted to fasten the panels in a way that would keep the gaps between the boards constant and minimize any distortion, such as cupping, due to seasonal changes in moisture content. To do this, I glued the boards along the outer edges of the door on their outer edges, and the others with a narrow bead of glue along their centers. I glued one board at a time, starting from one edge, and used cardboard spacers between the boards during assembly to create space for expansion. The unglued splines allow for movement.

The door is most easily finished with it lying flat on a table. I stained the outside of this door and then gave all exposed surfaces three coats of polyurethane varnish.

This door is 3½ in. thick overall. Its feeling of mass and sturdiness goes well with the post-and-beam construction. The total R-value is 7.25, calculated by the methods in the USDA Forest Products Lab's *Wood Handbook* ($1 from the U.S. Government Printing Office, Washington, D.C. 20402). By comparison, a standard 1¾-in. solid-core exterior door has an R-rating of about 2.5, and double-glazed windows have an R-value of about 1.9.

Irwin Post is a forest engineer. He lives in Bernard, Vt.

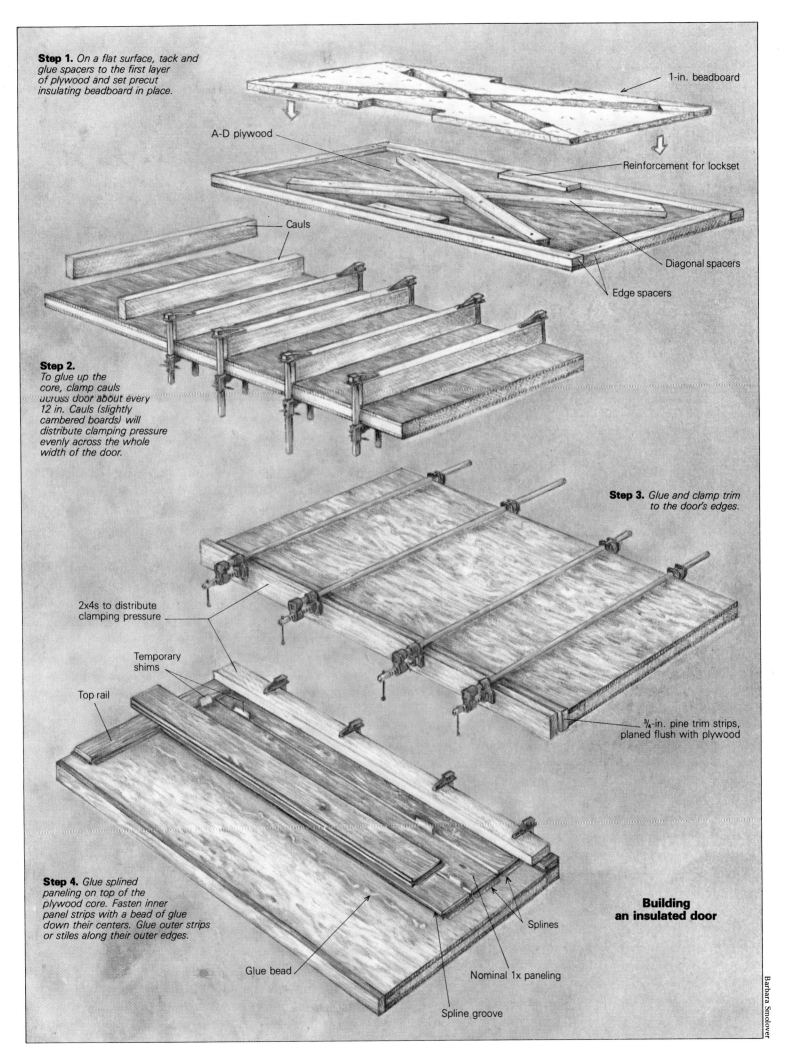

Batten Doors
Building a solid door from common lumber

by Bruce Gordon

A fine-looking batten door can be made from materials sold at any building-supply house, and can be built with limited funds and equipment. In the years when our business had no shop and little machinery, we produced custom batten doors at job sites, using only a table saw, an electric drill and a few clamps. They looked great, and were also competitive in price with factory-made doors.

Batten doors do have some inherent problems, though. Wood moves. A 36-in. door can vary as much as 3/8 in. in width between a dry winter and a humid summer. This will show on the side opposite the hinges, and the door that fits perfectly this winter may need to be planed down next summer and have its latch mortise reworked. The problem can be minimized by accommodating wood movement in the construction. Sealing the wood also helps, but if you use an oil finish, the door will move more than if you use varnish.

As a rule, batten doors do not stay perfectly flat and straight. They tend to bow across their widths and sag away from their hinges. The severity of these problems will depend on the species, grading, dryness and thickness of the lumber that you use, and how carefully you put the door together.

Boards 1 in. thick are best for interior doors, as are 1 5/8-in. boards for exterior doors, although you can use 3/4-in. tongue-and-groove stock for interior doors and 1 1/2-in. stock for exterior doors. The batten should be 1 1/4 times as thick as the door body, and 6 in. to 8 in. wide.

To begin, select the stock and cut it to approximate size. If it is roughsawn, joint one face, thickness-plane, joint one edge, rip to width, cut to length, tongue or groove (or half-lap) each edge and do any decorative milling.

I edge-join the boards with dowels to keep the door from sagging away from its hinges and use a Stanley self-centering doweling jig to drill two holes for 2-in. long hardwood dowels in the mating edges of all the boards. The holes should be drilled level with the eventual position of the hinges—about 13 in. from the top of the door and 7 in. from its bottom. Dowel diameters vary with the thickness of the stock, but I use 3/8-in. dowels with 3/4-in. boards, and 1/2-in. dowels with thicker stock. Once the holes are drilled, I apply a sealer (usually Watco or tung oil) and a first coat of finish or stain to the boards. On a door that will be painted, a coat of primer will do.

On a flat surface, assemble the door, inserting the hardwood dowels in the drilled holes. The dowels should not be glued, nor should the boards be pulled up tightly in the clamps. Instead, I insert strips of Formica between the edges to produce uniform gaps between the boards (photo facing page, top). The resulting gaps allow the wood to expand.

Being careful to keep the eventual location of hardware in mind, lay out the battens on the back of the door. Several pattern possibilities are shown in the drawing below. Check to be sure the door is square and flat, then clamp the battens in place. Battens should never be glued to the body. Attach them with metal fasteners so the wood can expand and contract freely.

I use four types of fasteners: rose-head clinch nails for their old-style look and ease of application, drywall screws for speed when I'm not concerned with the looks of the batten side of the door, wood screws, countersunk or counter- bored and plugged, and carriage bolts for the substantial look they give the face of the door. In any case, the batten should have an oversized hole, to allow the body of the door to move without bowing the door. Be careful to predrill even for clinch nails to avoid splitting the wood where the nail breaks through on the opposite side. Ideally, the clinch nail should be bent twice so that it penetrates back into the wood (drawing, facing page), ensuring a tight fit. However, it is common practice simply to fold the clinch nail flat. When the battens are secure, remove the Formica spacer strips.

A few tips regarding hardware may help you avoid frustration. Interior batten doors are usually too thin to take either a mortised latchset or a cylinder latchset. Consequently, you should plan on either a thumblatch or a rim lock. Batten doors are most often installed with a strap hinge, an H-L hinge or an H hinge.

If you don't want the traditional look such hinges give a door, you can use butt hinges. They should be sized so that the screws fasten to the edge of the door itself, not to the end grain of the battens where they won't hold. There are also offset hinges that can help you work around batten placements. Size the hinge so that its throw and the length of the batten allow the door to open 180° without hitting the casing. In some instances it may be best to hang the door from the casing, not the jamb, or use a half-surface hinge, one that combines a strap across the surface of the door with a butt plate mortised into the jamb. □

Bruce Gordon is a partner in Shelter Associates, a design and building firm in Free Union, Va.

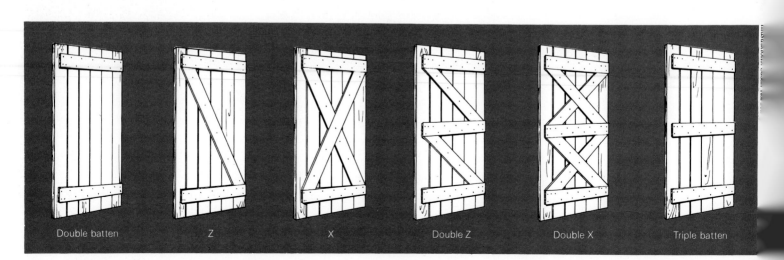

Double batten   Z   X   Double Z   Double X   Triple batten

Tongue-and-groove stock cut to approximate length is laid out across two sawhorses and clamped together flat and square. Two hardwood dowels inserted in holes in the mating edges of each board prevent sagging (detail, below). Above, battens are clamped to the back of the door. Formica strip spacers produce gaps that will permit the inevitable swelling. Battens should be fastened with either drywall screws, clinch nails or wood screws (left to right in bottom photo) or carriage bolts. Battens are predrilled for oversized holes, right, to allow for wood movement.

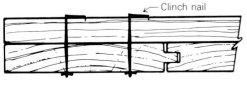

Clinch nail

Photos: Susan Mortell

Replacing an Oak Sill
Doing the job on a formal entry without tearing out jambs and trim

by Stephen Sewall

When I undertook the task of repairing the front entry of a Colonial Revival house in Portland, Maine, with its large 3½-ft. by 8-ft. door and side lights, I knew that the most difficult part would be replacing the rotted oak sill. It had suffered the neglect that many do, eventually checking and rotting from exposure to the weather because it hadn't been given periodic coats of sealant. The other repairs—which included replacing the raised panel in the door, replacing the pilaster bases, repairing the side lights and making some crown molding to replace part of the portico trim—were reasonably straightforward, but the sill presented some problems.

It seemed impractical to remove the entire jamb, replace the 7-ft. long sill and then reinstall it as the original (drawing, facing page) had been. Removing the jamb would require dismantling much of the trim inside and out, so it would be a big, labor-intensive job. Also, disturbing the entry that much could make refitting the door and side lights more difficult. Finally, I decided it would be best to replace the sill while leaving the jambs—both the inner jambs and the outer jambs at the side lights—intact.

Before the old sill could be removed, though, the door and the side lights had to be taken out. The side lights were held in place by stops on four sides. Since all of the trim was in good shape and I wanted to use it over, I was very careful to pry the stops out without damaging them.

The best tool I have found for removing any sort of wood trim is the Hyde #45600 Pry Bar-Nail Puller-Scraper. It has a thin blade that you can insert under almost any piece of trim without damaging either the molding or the surface to which it's attached. The thin end can also be sharpened so that you can cut small wire finish nails by hitting the other end of the bar with a hammer. The curved end of the pry bar can be used to open the joint up further and to scrape down the crusted paint before you reinstall the trim.

The stops of the side lights were inside mitered like most window trim, and because the side stops went in after the top and bottom, they had to be removed first. It's best to start prying at the middle, because the miters lock the ends in place. On each side light, I had to

---

Stephen Sewall is an architectural woodworker in Portland, Maine. Photos by the author.

The author installed a new sill in this Colonial Revival entryway before going on to replace the raised panel in the door and repair the side lights and part of the portico trim.

cut a few of the nails near the ends of the stop with the pry bar.

As soon as the stops were removed, I marked their back sides so I'd be able to put them back where they belong. I pulled the remaining nails through from the back so I wouldn't disturb the finish side of the wood. The tools I have found most useful for this are a pair of end snips or end pliers. They shouldn't be too sharp or they will cut the nail instead of pulling it out.

After removing the side lights, I cut plywood panels to fit between the jambs. These would keep the jambs from floating free when I removed the sill. I made the panels short enough so that I'd have room to cut out the old sill underneath them (photo facing page). I attached the panels to the inner and outer jambs with drywall screws.

With the door and side lights removed and the jambs locked in place, I was ready to cut out the sill. I used a Sawzall with a 6-in. blade to make cuts through the sill on either side of both inner jambs, and as close to them as possible (drawing, facing page). I made two more cuts 3 in. from each end of the sill, being careful not to hit the nails coming from the outer jambs into the end grain of the sill. This let me remove three large chunks of sill, leaving only the four pieces directly under the jambs.

I split out the remaining sections piece by piece with a 2-in. chisel. With all of the wood removed, I was left with 20d nails protruding from the end of the jambs. To get rid of these, I used a metalcutting blade in the Sawzall where I could, and a hand-held hacksaw blade on the less accessible nails.

The old sill had rested on five equally spaced 1-in. strips of wood embedded in mortar and running the width of the sill. I removed them and the mortar so I would be able to slip the new sill under the tenons of the door jambs. I also chipped away the mortar line between the brick and the old sill so that new mortar could be worked in under the outside edge of the new sill once I shimmed it into place.

The new sill—I saved the best section of old sill as a pattern for the new one. After a clean, square cut on the radial-arm saw, I was able to trace a full-size end profile. But even if I could have found a 14½-in. wide piece of 10/4 stock, the replacement sill would have been too hard to make in one piece. The 2-in. raised section under the door meant ¼ in. of stock would have to be removed from the rest of the sill. Making the sill in two sections, one ¼ in. thicker than the other, and splining them together would reduce the work—and the waste.

With the widths of 10/4 oak stock I had available, I made up the sill out of a 6-in. wide piece and an 8½-in. wide piece. The 6-in. section became the part over which the door would sit. This meant that I had only about a 3-in. width over which I'd have to waste ¼ in. of stock. After face-jointing and planing the stock to a net thickness of 2¼ in., I rabbeted out the ¼-in. by 3-in. section on the jointer. (If you haven't got access to a jointer, you could do this on a 10-in. table saw with the blade fully extended.)

The flat, raised section under the door was beveled at 3° to the back of the stock on the table saw. I cut a 45° bevel from the flat under the door to the point at which the finish floor contacts the sill. I used a rabbet plane to cut the small bevel on the other side of the flat.

The 8½-in. wide section needed its front edge ripped at 93° from the face so that it would be plumb when the sill, which would slope slightly toward the outside, was installed. I routed the top front edge with a ba bearing rounding-over bit. A ⅛-in. by ⅛-in. drip kerf was cut under the front edge of this piece to keep water from finding its way under the sill.

With all of the bevels cut and the front rounded over, I made the cuts for the splines. I cut a ½-in. slot 1 in. deep on each section of the sill, with the top faces held against the fence of the table saw. I epoxied in a piece of Baltic birch for the spline.

The visible parts of the sill needed to be belt-sanded before the wood could be finished. It's worthwhile to scrape off excess epoxy squeeze-out while you're gluing up because it can dull sanding belts in a hurry once it's dry. I find that most putty knives spread the glue on the surface rather than pick it up. The flexible tip of a small artist's paint knife, available at most art-supply stores, works much better.

To finish the sill, I wanted to use something more durable than spar varnish with an ultraviolet filter. In checking our local marine supply, I found that I had a choice of two two-part polyurethane varnishes that are used on the topsides of boats—Petit Durathane (Petit Paint Co., Inc., Borough of Rockaway, N. J. 07866), and Interlux Polythane Super Gloss (International Paint Co., 2270 Morris Ave., Union, N. J. 07083). It is important to buy the thinner recommended for these products. The first coat needs to be thinned down, because the unthinned varnish is a thick syrup that won't penetrate sufficiently. Be sure to use these products in a well-ventilated area. They smell terrible, and the vapors aren't especially good for your health. I applied three coats.

Installing the new sill—The top of the pilaster bases were in the way of installing the sill, and they also needed to be replaced, so I removed them. The bottom of the pilaster itself was also in the way, but the distance between the pilasters was only ¾ in. less than the finish length of the sill. I used my Japanese *azebiki* saw (it has rip teeth on one side, crosscut teeth on the other) to cut small sections out of the pilasters. Its thin, flexible blade makes a clean cut with a narrow kerf. I set the pieces aside to be epoxied back on once the job was complete and the sill and pilaster bases were in place.

I cut the new sill to length with a skillsaw against a homemade guide clamped onto the sill. To fit around the outer jamb framing, I cut 2-in. by 8-in. notches out of the inside ends of the sill. These were cut partway with the skillsaw and then finished off with a handsaw. With these cuts made, the sill could be slid into place.

My new sill was a little thinner than the old one, and the old mortar had been cleared away, so there was room for the sill to slide in under the tenons of the door jambs. With the sill held up snug against the tenons, I marked the mortise locations on the sill. Before I pulled it back out, I took a rough measurement at the back edge to see how much it would have to come up to be at the correct height to the finish floor. This gave me a di-

Doorway with original sill

The original sill was set on wood strips bedded in mortar on top of the brick base.

To start the job, Sewall removed the doorway's side lights and nailed plywood panels in place, with clearance beneath them to slide in the new sill. Then he removed the original sill, which had been resting on 1-in. strips of wood set in a bed of mortar.

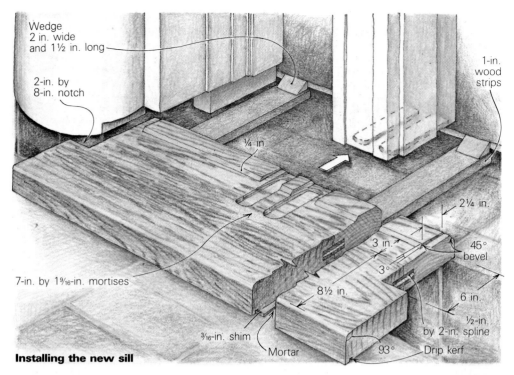

Installing the new sill

The new sill was splined together from two pieces of 10/4 oak, cut to the dimensions and angles shown in the drawing. The double mortises in the sill received the tenons at the bottom of the door jambs (photo left). Once the sill was in place, the pilaster bases were reinstalled (photo right), using shims to close up the joints between the middle and top half-round sections. The slivers that were cut out of the pilaster so that the sill could slide in were later glued back in place.

sill's installation. As the sill rode up on the wedges and came tight to the jambs, the excess butyl and epoxy squeezed out and had to be cleaned off. I reinserted my four shims under the front edge, two at each end and two at the door jambs to lock the sill into place.

During all this, the door jambs were held in position sideways by the plywood in the sidelight openings, but they could move in and out slightly in the sill mortises. My next task was to make them plumb so that the front door would close properly. I ran a string on the inside from one outside jamb to the other, then tapped the door jambs into position with a block and hammer. To hold them there, I toe-screwed through them into the sill with 3-in. screws. I also put several toe-screws into the sill at the outside jambs because there was nothing else there to prevent the sill from creeping out, except the cedar shingles and the mortar under the front edge.

I mixed the mortar in a loose batch and worked it under the front edge of the sill. I snapped off the cedar shingles and knocked them in far enough so they would be covered by mortar. With a pointing tool, I worked as much mortar as possible back under the sill. I cleaned up the excess mortar with a stiff brush and water. The butyl caulk was scraped off, and the surface was wiped down with a rag soaked in paint thinner.

Finishing up—The old pilaster bases had been saved for samples so that new ones could be faceplate-turned on the lathe. I used mahogany because it is available in large dimensions and is more resistant to rot than pine. There were three pieces to each base: the square bottom, a half-round middle section and a scotia and half-round on top.

Before I installed the bases, I primed them with oil-base paint. To cushion the wood and to keep the square base from touching the granite below, I covered the base's bottom with butyl caulk and a layer of lead, which was fastened with copper nails.

The top section of the base fit into a rabbet at the bottom of the pilaster, so it could not be slid into place. I caulked the pilaster with butyl and held the upper section in place. With the square section put in place, the middle half round could be slid between the two.

To tighten up the three base sections, I inserted cedar shingles between the bottom two (photo above center). At some points on the circle no shimming was needed, and at others as much as 1/8 in. was needed. I used 8d galvanized finish nails to pin the sections to each other and the pilaster to the base. The cedar shingles were cut off as far in as possible with a utility knife, and the joint was caulked with butyl. Finally I epoxied and nailed back the slivers that had been cut out of the pilasters so the sill could be slipped between them.

I hung the front door as soon as the epoxy glue in the door jambs had hardened and the mortar under the sill had set. The door fit just as it had when I removed it. The side lights, stops and plinth blocks were reinstalled, and the job was done. □

mension for the wedges needed to lift the back edge snug into place.

I used a router with a 1/2-in. straight-face bit to make the four mortises in the sill. I cut the mortises freehand, taking successively deeper cuts until the depth was just over 1/2 in. Each inner jamb had a double tenon. I cut each mortise 1/8 in. wider than the tenons and extended the mortises through the inside edge (drawing, and photo above left). The extra length was necessary because the wedges would start to push the mortise onto the tenon before the sill was all the way back into place. The extra length would be covered up when the inside plinth block was reinstalled.

I made two wedges out of oak to use for the back edge of the sill. They were 2 in. wide and 1½ in. long. I calculated one to be 5/8 in. high and the other 3/4 in. These were set on the subfloor under where the sill would rest when it was slid into place. I used a hand plane to put a slight 45° chamfer under the back edge of the sill so it would slide up more easily on the wedges.

On the second dry fit, the sill slid up the wedges and I got a pretty good fit against the jambs, but one of the wedges needed to be changed to get a perfect fit. With the two wedges adjusted for height, I was ready to make three more to support the entire back edge of the sill. I ran a line from the top of the two wedges and took the measurements at the intermediate locations. With all five wedges in place, I tried a final dry fit. This time I used cedar shingles to shim up the front edge, where the mortar would eventually be packed. I needed to shim up the front edge only about 3/16 in.—just right for a good mortar joint.

Before installing the sill, I spread epoxy on the tenons and in the mortises. I also ran a bead of butyl caulk underneath all of the shoulders of the jambs, and spread a stiff batch of mortar between the wedges and as high as I could without interfering with the

A Breath of Fresh Air

Making a Victorian screen door

by Alasdair G. B. Wallace

On a trip through southern Ontario, I was impressed by the number of carefully restored and lovingly maintained older homes. Wherever I looked, there were beautiful old doors—some were hand-carved, many swung on wrought-iron hinges, others had leaded lites. But it saddened me to see these original doors hiding behind standard aluminum combination storm-and-screen doors that lacked aesthetic rapport with their surroundings. Southern Ontario does require storm doors and screen doors in the appropriate seasons. Antique doors occasionally show up at country auctions, but they usually command exorbitant prices. Making such a door, however, requires only basic woodworking skills, and will provide you with a good excuse for an indoor project this winter and a welcome breeze next summer as well.

The door detailed in this article is a copy of a well-worn original. It consists of an inner oval frame enhanced by turned spindles, a separate lower decorative screened opening, and a raised panel at the base (photo right). The frame of the door may be joined by dowels (as was the original) or mortised and tenoned. I recommend the latter for its greater strength. You'll require a table saw, basic woodworking hand tools and a lathe. A bandsaw or saber saw will speed some of the work. Some of the cuts and grooves described below can also be made with a radial-arm saw, a shaper or a router.

Materials and hardware—Regardless of the wood you use—oak and pine are popular in Ontario—you'll need boards that dress out to a minimum thickness of ⅞ in.; 1 in. is preferable. Anything less will result in a flimsy door. Select wood that is dry (9% moisture content or less), straight grained, and quarter-sawn if possible. You'll need about 18 bd. ft. of stock, and about 2 bd. ft. of 4/4 maple for the spindles.

For my door I purchased new hardware from Lee Valley Tools Ltd. (2680 Queensview Drive, Ottawa, Ont., Canada K2B 8H6). You might also take a look at their catalog of antique hardware. You'll need one 9-in. Chicago door spring with an adjustable tensioning device and a cast-iron screen-door set, which includes a mortise lock, knob, drop handle and strike plates. Make sure it is suitable for doors at least 1 in. thick. Brass sets are also available. Also get three standard butt hinges. Since I planned on alternating the screen door with a standard storm door as the seasons changed, the hinge leaves for each are identical in size and placement, and align with the leaves on the door frame. I have only to remove the pins to change doors.

For screening, you'll need enough to allow a 2-in. overlap on all sides of the opening. Bronze and copper screening (if you are able to obtain them) are strong, an important consideration if you have children or a pet or are plagued with squirrels. I used black anodized aluminum screening. It is easily installed and becomes almost invisible against the shadows.

Laying out the plans—Don't assume that your door frame is either square or plumb. Measure from corner to corner diagonally. If the diagonals are the same, your frame is true; otherwise you'll have to make allowances in the door. Measure the width and height at several locations, then make your screen door ¼ in. shorter and ⅛ in. narrower than the opening. In calculating the length of your rails, remember that you need to add 3 in. for the 1½-in. long tenons (drawing, next page).

Making the door—Select the straightest-grained boards for the stiles. Once you have marked out your stock, rip it to the required dimensions and square the edges. I prefer to remove any planer marks with a sharp smoothing plane because it leaves the surface of the wood bright and crisp-looking. If you wish, though, you can sand them out. To avoid possible error later, lightly pencil in the future location of each piece—top rail, left stile—on its outdoor face.

The design of this door includes a decorative molding, which may readily be cut on the rails and stiles with a shaper head such as a Craftsman 9-2352 AM or a Multiplane. I prefer the Stanley 45

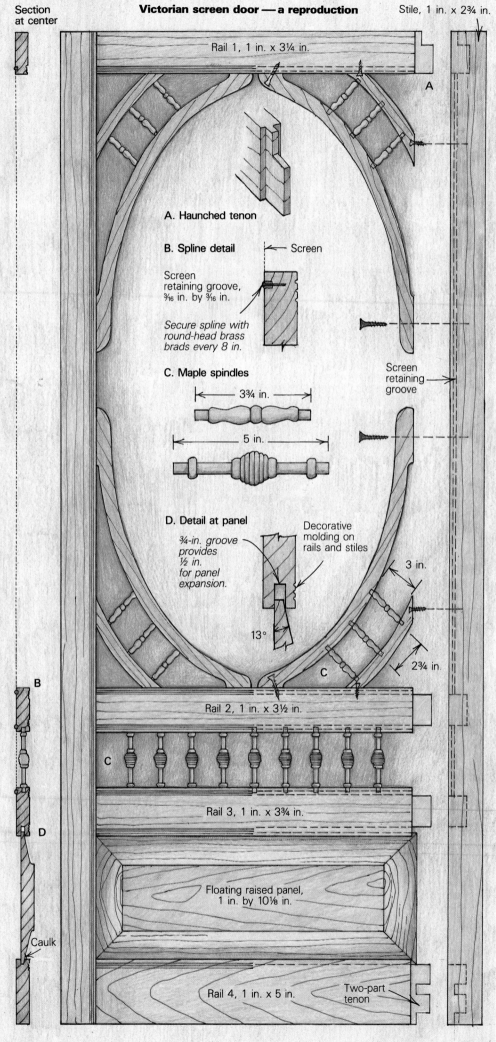

grooving plane, but this old tool is hard to come by. A series of passes over the angled blade of a table-saw will achieve a similar effect, or you could use a router. But to avoid weakening either the mortises or the groove for the panel, run the pattern to a depth of only 1/16 in.

The tenons should be approximately one-third the thickness of, and centered in, the stock. For additional strength, tenons in the top and bottom rails should be haunched (see detail A at left). Because the bottom rail is 5 in. wide, its tenon should be in two parts in order to reduce the possibility of weakening the stile.

Once the tenons have been marked out, they may be cut in the traditional manner with a backsaw. An alternative method uses the simple tenon-cutting jig described in the sidebar on the facing page.

Mortises can be simply marked out by laying the rails across the stiles in the appropriate location and marking the edge of the stock around the already cut tenons. Mortises may be cut with a router, but you can also chop them out with a mortising chisel. If you use a router, either the edges of the tenons must be rounded over with a rasp to match the rounded corners of the mortise, or the mortise must be chiseled square to match the tenons. Either way, accuracy is essential if you want to avoid having the door twist out of shape.

Run the grooves for the lower panel in the noted locations in rails and stiles. Making these grooves the same thickness as the tenons and centering them on the edge of the stock enables you to run them through to the ends of the stiles, where they serve to locate the haunched part of the tenons. If you're using the dado head on a table saw, be careful not to over-run. Mark the location of the blind end of the groove in pencil on the face of the stiles. Mark the location of the leading edge of the dado cutter on the rip fence, stopping the groove just before the two marks align. Square up the stopped ends with a chisel.

After you have dry-assembled the frame of your door to check dimensions and flatness, mark on the inside face of the door the location of the 3/16-in. by 3/16-in. groove that will house the screen-retaining splines. This groove should be 1/4 in. from the inner edge of rails 1, 2, and 3 and the stiles. Running these grooves the length of the stiles will considerably weaken the door in the area of the mortises and the panel-retaining groove. Instead, I recommend that you match the ends of these grooves with stopped grooves in the stiles. A router is a most convenient tool for this operation.

To make the panel, you'll have to glue up several narrower boards. Make sure you alternate the direction of the annual rings in order to minimize the liability of warping. Rather than shape both faces of the panel, I decided to leave the back flat. This results in the front face protruding beyond the rails and stiles, adding a bolder relief to the bottom part of the door.

Once the panel blank has been dimensioned, the center raised portion can be formed on the table saw. To do this, tilt the blade to an angle of about 13° and run the panel through on edge, using a 6-in. high rip fence to help support the work. With a sharp carbide-tipped rip blade, you can raise the panels in a single pass. To cut the small vertical shoulders that border on the field, lay the panel face down and use the fence as a guide. Experiment with scrap wood to achieve the effect you want. Clean up the bevels with a rabbet plane and sandpaper block. The finished panel edges should be ¼ in. thick.

The width across the grain of a 10⅛-in. panel may vary considerably, depending on the species of wood and the ambient humidity. To accommodate this potential movement, run the panel-retaining groove in the bottom of rail 4 to a depth of ¼ in., and allow about ¾ in. for the groove in rail 3. The panel will be fully seated in the first groove, and the second will accommodate expansion.

Spindles and inserts—I experimented with several shapes and spacings of spindles before deciding on the combination that looked best. I used nine of the smaller spindles and twelve of the longer ones (detail C on the facing page). Mark the locations of the spindles on the bottom of rail 2 and the top of rail 3, and center the holes in the stock unless your frame is less than 1 in. thick. In that case, offset the location of the holes toward the outer face of the door so that they don't run through to the screen-retaining groove on the inner face of these rails.

The inserts that define the oval shape of the door's interior also serve to strengthen the frame. Prepare two full-size patterns for the two different parts of the oval frame insert, and lay out the stock so that the grain runs with the length of the stock. After the stock has been marked, cut out the inserts with a bandsaw or saber saw. On my door, I rounded the inner edges slightly with a ¼-in. cove bit fitted to my router. Use a framing square to make sure the ends are square to each other. The inspiration for this screen door had inserts that were nailed in place, but I prefer to fasten the inserts with 2-in. #9 brass screws, countersunk.

Once the entire door (with the exception of the oval inserts and spindles) has been dry assembled, test-fit it in the door frame. Note and correct any irregularities before beginning the finishing process.

Finishing and assembly—Traditionally, many pine doors were painted, while others were finished with a varnish stain and still others were grained to simulate oak. The door I built was lightly stained to match the existing pine exterior door, then finished with four coats of exterior satin Varathane. I rubbed each coat down with steel wool, using progressively finer grades down to #000.

I learned the hard way on the first door I built that finishing is much easier if it is done before the frame is glued up. This makes it easier to get at all the nooks and crannies of the completed door, and later, when the raised panel shrinks during dry weather, no unfinished margins will appear. Finish the spindles as they rotate on the lathe; then cut them to their final length. Be careful, however, to keep finish away from any portion of the door that will receive glue.

Gluing up will proceed more smoothly and rapidly if you have a helper to assist in aligning the panel and the lower spindles. Check that the frame is flat and square. If you need to adjust it, a long clamp placed diagonally from corner to corner will enable you to squeeze it slightly into square. (An additional advantage of finishing the parts of the door prior to assembly is that any excess glue can be readily removed from the finished surface.)

Don't glue the panel in place. Instead, leave it to float within the grooves in the surrounding frame. If your door will be exposed to the weather, run a thin bead of clear silicone caulking along the exterior seam between panel and bottom rail to keep water from collecting in the groove. Once the glue has set (remember to use a waterproof plastic-resin glue), the oval inserts and spindles may be installed.

Installing the screens—The next step is cutting enough ³⁄₁₆-in. by ¼-in. pine splines to retain the two separate screens of the door. I gently rounded the top edge and tapered the sides of my splines with a block plane. The taper accommodates the double thickness of screening within the groove, and the pine itself is soft enough to compress slightly in the groove.

You don't need anything fancy to persuade the screening into place. I used a scrap of metal plate and lightly tapped it with a mallet to push the screen into the grooves, long sides first. Enlist your helper to stretch the screening as you urge it into the grooves, checking frequently that it is taut. For a good-looking job, make sure that the weave of the screen aligns with the frame. Cut your splines to length and tap them into position, using a scrap of wood to avoid scarring them. Secure the splines every 8 in. with ¾-in. brass escutcheon pins. Any excess screening along the edges can be removed with a razor knife run along the outer edge of the spline—I trim it flush with the door. Finish the splines to match the door before installing them.

Installing hardware—Test-fit your door and mark the location of hinges and the handle. Then install them and hang the door. On my door, I installed a Chicago door-closing spring with adjustable tension. To eliminate most of the racket when the door bangs shut, I slightly recessed three 2-in. lengths of self-adhesive weatherstripping in the door frame.

I'm a slow worker, so this door and another one I made took me quite a while. But the breeze from the verandah through the kitchen area and the open feeling on the back porch make all the work worthwhile. My neighbor stopped me as I was mowing the lawn this morning, asking if I might find time to make a door for him. But I think I'll try designing one for our front door first, in maple, maybe, to match the staircase. □

Alasdair G. B. Wallace, of Lakefield, Ont., is a contributing editor of Fine Homebuilding. *Photo by the author. For more on this subject, see Amy Zaffarano Rowland's* Handcrafted Doors and Windows *(Rodale Press Inc., 33 E. Minor St., Emmaus, Pa. 18049, 1982; $12.95 paperback; $21.95 hardcover).*

A useful tenoning jig

If you have a series of tenons to cut, this tenoning jig is simple, inexpensive and will save you a lot of time—with one pass over your table saw you can cut both cheeks of a tenon (drawing, below). The idea is not mine, but I've been using it for years and am indebted to whoever figured it out.

You'll need two identical blades. I use 9-in. planer blades, running side by side in my 10-in. table saw. I use these blades only for tenoning, and sharpen them together in order to keep their diameters identical. They are separated on the saw's arbor by a metal spacer. A local machine shop made me a set of spacers, ¼ in., ⁵⁄₁₆ in., ⅜ in., and ½ in. wide, and they have saved me endless hours and plenty of frustration. Check your saw's throat-plate clearance before you order your blades. To accommodate the width of the paired blades, I had to make a special table plate out of plywood. For safety, the throat-plate opening should be only slightly larger than needed to clear the blades.

The jig itself consists of a U-shaped wooden device which fits over the rip fence and slides along it. The fit should allow the jig to slide freely but not be so sloppy that

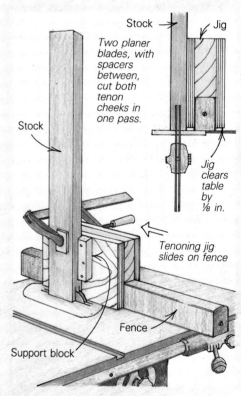

the workpiece wobbles. To minimize binding, you might try waxing the inside of the jig, or the fence. To cut the tenons, install the appropriate spacer between the blades so they will cut the exact thickness of tenon you wish. Raise the blades a distance equal to the length of the tenon, and adjust the fence with its sliding jig in order to center the tenon in the stock. Align the stock against the vertical backing block, clamp it in place, run it through the blades and you have two perfect cuts in one safe, efficient operation.

To remove the waste and cut the shoulders, make up an end-stopping attachment for your saw's miter gauge. Set your miter gauge at exactly 90° to the saw blade, and use a single planer blade to remove the waste. —*A. G. B. W.*

Building Wooden Screen Doors

Durability and aesthetics hinge on sound joinery and steadfast materials

by Stephen Sewall

According to Steven J. Phillips' Old-House Dictionary, the purpose of a screen door is "to allow ventilation but exclude insects." Of course if it can accomplish that gracefully, all the better. During the late 1900s, when screen doors first became popular in this country, most were designed to enrich, or at least complement, the architecture. To that end, a wealth of styles emerged, many of which were featured in pattern books and copied by carpenters. Even many of the mass-produced screen doors in those days were both fetching and functional.

Nowadays, though, the ubiquitous aluminum screen door is the one most likely to appear on someone's doorstep. To be sure, aluminum doors have their place, but all too often they wind up in classical entryways or over finely wrought entry doors, spoiling the intended effect.

A wooden screen door, on the other hand, can be unobtrusive, allowing the main door to show through. Or it can mask the door behind it and serve as an architectural ornament in its own right (photos facing page). As a homebuilder and architectural woodworker, I've built screen doors both ways. And in either case, the basic construction principles are the same.

Screen-door anatomy—In its simplest form, a screen door consists of two stiles, a top rail and a bottom (or kick) rail (drawings, p. 50). Traditionally, most screen doors were built of pine, oak and other domestic woods. But pine is soft, and oak isn't especially stable or weather resistant. Cypress is a good choice, but here in Maine it's hard to find. Though I've used a variety of woods for my screen doors, I prefer Honduras mahogany. It's strong, stable and holds paint and varnish well.

I build my doors 1⅛ in. thick, a compromise between light weight and sturdy construction. Originally, I joined stiles and rails with dowels because it was quick and easy. But after a while, the joints loosened. Screen doors bear the brunt of the weather and they get slammed a lot, especially when they're spring-loaded. Pneumatic closers (the kind you see on most aluminum screen doors) are especially hard on wooden screen doors because they hesitate in mid-swing and cause excessive racking. Nowadays, I use mortise-and-tenon joinery exclusively for my screen doors.

Stiles and top rail are a minimum of 4-in. wide, which makes for a sturdy door and provides plenty of room for locksets and springs. The kick rail is at least 8-in. wide to keep the door from distorting. I cut two narrow tenons instead of one wide tenon in the kick rail so that the mortises don't weaken the stile.

Tenons are 3 in. long and ⅜ in. thick, or one third the thickness of the stock. A ½-in. wide by ⅜-in. deep rabbet along the inside edge of the frame accommodates the screen and screen stop. That's all there is to it.

Making the door—One approach for making this type of a door is to cut the mortises and tenons and glue up the door first, and then rabbet and groove its inside edge with a router to accept the screen and stop (this requires that the rabbets be chiseled square at the corners). But I prefer to rabbet the stiles and rails *before* assembly, extending the rabbets down the full lengths of the stiles. This requires that the ends of the rails be cut to engage the edges of the stiles (top right drawing, p. 50). The cuts are easily made with dado blades on a table saw, and the resulting lap joints strengthen the door.

I start by selecting clear, straight-grained stock with a moisture content of about 10%. After sizing and jointing the stock, I lay out the mortises on the stiles. I stop the bottom mortise about 1-in. short of the bottom end of the stile so that the door can be trimmed without cutting into the mortise and tenon. One inch of stock separates the two mortises for the bottom rail. On the top end, I hold the mortise back about ½ in. On the inboard sides, the mortises stop where the rabbets begin.

I cut the mortises with a horizontal boring machine. I prefer this over a hollow-chisel mortiser in a drill press because the boring machine is faster and more accurate. The only drawback is that it produces mortises with rounded corners. To avoid squaring the ends of the mortises or rounding over the edges of the tenons, I cut the tenons narrower than the mortises by the diameter of the mortising bit, or ⅜ in. That leaves small half-circle hollows on either side of the tenons, which fill up with glue during glueup. I also cut the short rabbet at the mouth of each mortise with the boring machine, squaring its inside corner with a chisel. Mortises can also be cut with a router or chopped out by hand with a mortising chisel.

I cut the tenons on a table saw fitted with a ¾-in. wide dado head set to depth of ⅜ in. I lay the stock flat on the table and push it through carefully with a miter gauge set to 90°. The rip fence on the table saw serves as a stop to determine the length of the tenon. Starting with the shoulder cut and pulling the stock away from the fence with successive cuts, I need to make only five or six passes to cut a 3-in. long tenon.

I readjust the fence to cut the other side of the tenon so that it's ½ in. longer than the first side to compensate for the screen rabbet. After the cheek cuts are completed, I adjust the height of the dado head, flip the rails on edge and trim the tenons to width using the same method. I cut out the space between the double tenons on the bottom rail with a bandsaw. Tenons can be cut a number of other ways, of course, such as with a single-end tenoner, a tenoning jig on a table saw (see pp. 45-47), a bandsaw or even by hand with a backsaw.

With the mortises and tenons completed, I cut all the rabbets using either the dado head on the table saw or a shaper, which gives a cleaner cut. The ½-in. width of the rabbet gives ample room to center a ⅛-in. wide by ³⁄₁₆-in. deep (approximately) groove. The screen will be pressed into this groove and held fast with ⅛-in. dowel stock (more on that later).

I cut this groove with a combination blade on the table saw (the inside corners are connected after glueup with a small chisel). The width of the groove is crucial and must be determined by trial and error with a piece of screen and dowel. The dowel must fit snugly and tightly enough to hold well, but not so tightly that the dowel can't be pressed below the surface of the wood. When the dowel fits correctly, the door is ready for glueup.

Gluing up—Before gluing up a door, I assemble it dry to make sure everything fits. I use two-part epoxy for my screen doors. Though it's expensive and has a relatively short pot life, it's waterproof, has excellent gap-filling ability and is transparent when it cures. Unlike most other glues, it actually bonds better on sawn surfaces than on planed wood. Also, epoxy can be mixed to be a bit

Wooden screen doors can be unobtrusive or ornamental, depending on the desired effect. The screen doors in the photos below are simple, allowing the entry doors to show through. The screen door pictured above, built of solid oak, conceals the main door and serves as an important visual element in the entryway.

The door below, built of Honduras mahogany, is simple and sturdy. The lock rail adds strength and divides the upper and lower half of the door behind it. The door pictured above is a stone's throw away from the ocean and features bronze screening, which resists corrosion. The storm doors over it protect it from nasty weather.

flexible so that it will move with the wood. Though epoxy usually requires a temperature of 65° F or above to cure, special formulas are available for use at colder temperatures. Epoxies have also been developed that cure quickly or adhere well to specific types of wood. The epoxy I use is made by Allied Resin Corporation (Weymouth Industrial Park, East Weymouth, Mass. 02189), but there are plenty of good brands available.

I use bar clamps and leave them on overnight. A long clamp tightened diagonally straightens the door if it's out of square.

If the door is to be finished, I finish it right after glueup, but before installing the screen. My doors are usually finished with an oil-base primer and paint, or with spar varnish with a UV filter.

Installing the screen—With the frame completed, the final step is installing the screen. There are several different types of screen on the market, though they aren't all easy to find. Most hardware stores carry aluminum and fiberglass screening, which are relatively inexpensive. Bright aluminum screening lets the most light through, but it dents, tears and corrodes easily. Charcoal-colored electro-alodized aluminum is tougher and resists corrosion. Fiberglass, the cheapest screen on the market, is easy to work with and won't dent. But it does stretch, and bluejays, grasshoppers and other critters like to chew on it.

I often use bronze screening. It's expensive and stiffer to work with than aluminum and

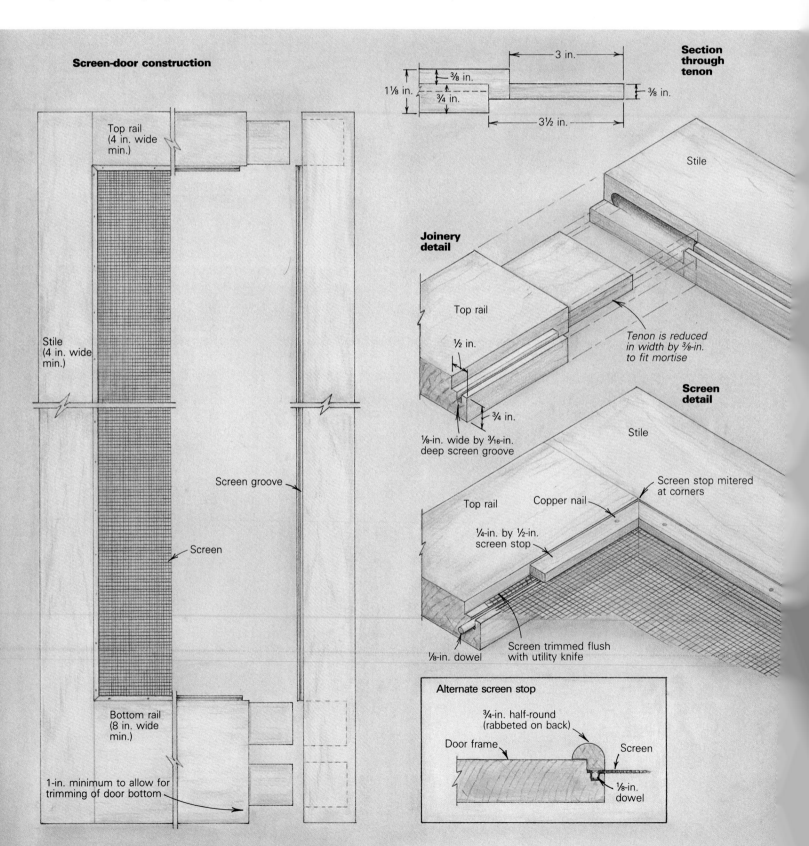

fiberglass, but it's strong and resistant to corrosion, an important consideration when installing screens in houses by the ocean (where I do most of my work). It also tarnishes to a greenish color, a look that some of my clients prefer. I buy mine at a local hardware store.

Screening is also made of galvanized steel (which turns chalky and disintegrates) and stainless steel (by far the most expensive and durable screen on the market). Some manufacturers even make a coated fiberglass screen called "solar shade" that blocks 70% of the sun's heat while providing ventilation. It's supposed to reduce air-conditioning bills and carpet fade. Hanover Wire Cloth, Inc. (E. Middle St., Hanover, Pa. 17331) calls their product Solar Guard or Solar View, depending on the mesh. The New York Wire Co. (152 N. Main St., Mt. Wolf, Pa. 17347) calls theirs Goldstrand Solar Screen.

Whatever the screening material, I use a splining tool to push it and the 1/8-in. dowel into the groove (photo below). A splining tool consists of a handle about 8-in. long with a narrow 1½-in. dia. metal wheel on either end. One wheel has a convex edge and is used to coax the screen into the groove. The other wheel has a concave edge and is used to press the dowel into the groove.

Because the tool is designed for installing screen in metal frames (where rubber gaskets are used instead of dowels to secure the screen), it doesn't fit the dowels quite right. But with practice and patience, it works fine. Some hardware stores sell splining tools, or they can be purchased from Elgar Products, Inc. (P. O. Box 22348, Cleveland, Ohio 44122).

I cut the screen 4 in. bigger than the opening, which leaves enough extra screen to grab onto and pull tight during installation. Starting at a stile, I press the screen into the groove with the convex wheel, making sure the weave of the screen is straight in relation to the frame. Then I turn the splining tool over and shove the dowel into the groove. This locks the screen into the groove.

Next, starting at the middle of the opposite side, I push the screen into the groove with the splining tool while pulling the screen tight with my other hand. The screen doesn't have to be perfectly tight because when I press the dowel into the groove, the screen is tightened further. After I install the second dowel, I repeat the process with the remaining two sides. When installed, the screen should be stretched nicely with no major depressions. If any objectionable dips remain, I pry out a dowel, tighten the screen and insert a new dowel.

The excess screen is bent into the corner of the rabbet with the splining tool and trimmed off at the corner with a razor knife. I cover the edge of the screen with a simple ¼-in. by ½-in. stop, mitered at the corners and fastened with copper nails (drawing facing page, lower right). For a more decorative effect, I sometimes use ¾-in. half-round instead, rabbeted on the back side so that it holds the screen and covers the edge of the door (drawing facing page, below right).

Hardware—I use solid-brass hardware for its durability. Stanley Hardware (a division of The Stanley Works, New Britain, Conn. 06050) makes a 3-in. by 3-in. stamped solid-brass hinge with a ball tip that is well-made, reasonably priced and looks good.

For the door latch, I use a surface-mounted lockset made by Merit Metal Products Corp. (242 Valley Rd., Warrington, Pa. 18976). The lockset has a knob on the exterior side, a latch on the interior side and is lockable. As a bonus, it's easy to install. I'm extra careful, though, to install the lockset where it won't bump into the lockset on the primary door (I learned that lesson the hard way).

Storm doors—There isn't much difference between a screen door and a storm door, except that storm doors are designed to inhibit, rather than encourage, air infiltration. That means substituting ¼-in. thick laminated safety glass or tempered glass for the screen. Some screen-door manufacturers build doors with interchangeable panels to fill both functions. But the doors require additional hardware to secure the panels. Also, it's difficult to stretch screen tightly over a narrow, removable frame without causing the frame to bow. Plus, large panels of glass are difficult to handle in removable frames. I prefer to make two separate doors that can be interchanged by removing the hinge pins.

The only difference between my screen doors and storm doors is that I adjust the size of the rabbets to accommodate the glazing, and of course, I eliminate the screen grooves. The glass is contained in the rabbet with either glazing points and glazing compound or with wood stops. When using glazing compound, I cut the rabbets ¼ in. wide by ⅝ in. deep, which allows the proper slope for the glazing compound. The same size rabbet also works well in conjunction with ¼-in. quarter-round molding. Otherwise I cut the rabbets ¼ in. wide by ⅜ in. deep and use ¾-in. dia. half-round molding. The molding is rabbeted on the backside so that it laps over the edge of the door frame. □

The screen is installed with the use of a splining tool. The tool consists of a handle with a metal wheel on either end, one with a convex edge and the other with a concave edge. To install the screen, the author presses the screen into the screen groove with the convex wheel. That done, he flips the tool over and presses a ⅛-in. dowel into the groove with the concave wheel (photo above), locking the screen into the groove. Excess screen is trimmed off at the corner of the rabbet with a utility knife.

Stephen Sewall of Sewall Associates, Inc., is a custom homebuilder in Portland, Maine. Photos by the author.

Production-Line Jamb Setting and Door Hanging

Time-saving techniques from the tracts

by Larry Haun

I went to work as a carpenter in 1949. A craftsman in white overalls taught me the trade I practice to this day, from foundation to finish work. In those early days, my crew and I would build two or three houses a year.

But post World War II America was booming. Good jobs with good wages were available to anyone who wanted to work. Veterans could take advantage of home loans under the G. I. Bill. Literally hundreds of thousands of people needed, wanted and could afford to own a new home. Instead of building one home at a time, carpenters needed to build 5,000 at a time, and as a result, we had to come up with more efficient construction methods.

And that's just what we did. It wasn't long before we were constructing a house in days instead of months. But even though these new methods were considerably faster than the old tried-and-true procedures, they proved to be just as effective.

Everything seemed to change. Hammers gave way to nail guns. Power tools became commonplace. And most important, housing production became an assembly-line process. Those who framed the walls no longer cut and stacked the roof. Those who set the jambs no longer hung the doors.

The new assembly-line methods generated an incredible increase in production. In 1950, we were expected to hang a door an hour and eight doors a day. But in 1953, a friend of mine, a door-hanging specialist, was hanging 80 to 120 doors a day with assistance from a helper. Not only was he fast, he was accurate; the quality of his work far exceeded that of carpenters who hung only a few doors a year.

Clipping the trimmers—Traditional methods of setting door jambs are quite effective, yet they require considerable measuring, sawing, nailing and shimming. A faster, but equally effective way to build door jambs is based on a technique called *clipping*. Clipping eliminates the need for shims—nails alone hold everything securely in place. Clipping even eliminates the need for door cripples (photo above), except for jambs in which pre-hung doors are being used.

A clipped-trimmer door jamb stands in the foreground of southern California's Local 409 instructional framing project. Note how the header is just below the top plate, eliminating the need for cripples. The trimmers have been plumbed, straightened and clipped to the king studs, and the jamb has been affixed to the trimmers without shims. The portion of the sole plate that used to be between the trimmers has been removed and installed as drywall backing above the head jamb.

Because you already know the stud length, there is no need to go from opening to opening, measuring each one. The length of each door header is equal to the width of the door plus 5 in. To set a door jamb using the clipping method, begin by pre-cutting all the necessary trimmers. Cut the trimmers about 1/16 in. oversize to ensure a snug fit. The next step is to put a trimmer under each end of every header. Halfway down its length, secure each trimmer to its king stud with a 16d nail.

Using a 6-ft. level, begin to plumb the trimmer. The level indicates which end of the trimmer doesn't have to move. Toenail that end into the header or bottom plate with one 8d nail in the center. No other face nail is necessary. Then use a straight-claw hammer to pull the un-nailed trimmer end out from the king stud. Once the trimmer is plumb, toenail this end with 8d nails.

Even though the trimmer is now roughly plumb, it will probably have a bow in it. Now it's time to straighten the trimmer (this part of the job eliminates shims). Hold the 6-ft. level on the trimmer (top left photo, facing page) and use the hammer's claws to lever the trimmer away from the king stud so that it's flush with the edge of the level. The 16d nail in the center of the trimmer temporarily holds it straight.

At this point, clip the trimmer to the king stud. Begin by driving a 6d or 8d nail partway into either the trimmer or king stud. Bend this nail back onto the other upright. Then drive and bend a second nail over the head of the first (detail photo, facing page). An experienced jambsetter can drive and bend this nail with one swing. Install three clips per side. This holds the trimmer true for the life of the building. Some jambsetters use a wide staple in place of regular nails to tie the trimmer to the king stud. This is a good method that further simplifies the process.

Use a "spreader gauge" to plumb the second trimmer. Assemble a door jamb, measure the width at the head from outside to outside and cut a 1x4 to this length. Place this spreader gauge up against the already-plumbed trimmer at the top and pull the unset trimmer against the gauge. Secure the second trimmer to the header with 8d nails. Do the same at the bottom. Then straighten and clip this trimmer just like you did the first one.

Now you are ready to cut out the bottom plate. To do this I use a worm-drive saw fitted with an arbor extension that allows me to make a cut that's flush with the saw's base. The one I've got is called a Close Cut, and it costs about $35 (Western Saw Inc., 1842 West Washington Blvd., Los Angeles, Calif., 90007). The Close Cut has a guard on top,

Trimmers are fastened with the help of a single 16d nail connecting the trimmer and the king stud (top left photo). Here the author levers the two members apart slightly while keeping an eye on the level bubble and the gap between the trimmer and the level. The friction of the nail will hold the trimmer and king stud in the correct relationship while the two are clipped together. Clamping a level to the head jamb allows you the freedom to hold a ½-in. thick block along the edge of the side jamb, approximating the wall-sheathing thickness (photo above). A pair of 8d nails that have been bent over to clip the trimmer and the king stud can be seen in the photo at left.

but it lacks the retractable guard common to circular saws. Because of this, I use it only in flush-cutting situations where the direction of cut is down and away, and I'm careful never to set the tool down while the blade is still moving. If you haven't got a saw fitted with one of these, a chainsaw or a reciprocating saw will do the job. Set aside the cut-out piece for later use as drywall backing.

Installing the jamb—Pick up the assembled jamb and insert it into the framed opening, making sure that it fits snugly. The side jambs will eventually be nailed directly to the trimmers, with no need for shims.

To level the head jamb, use a short level and clamp it to the jamb (top right photo). This frees your hands to do other things. Now use a small block of wood to represent the thickness of the wall covering. Typically, the block is a piece of ½-in. plywood that represents the most common wall covering—½-in. drywall. Place the gauge against the trimmer in order to hold the jamb out the required distance from the face of the wall.

Next, working from bottom to top on the trimmer, nail in five pairs of 6d finish nails into one side jamb. Keeping the pairs of nails separated by 3 in. prevents the jamb from cupping. Larger nails need to be used in jambs holding heavy doors.

Now check the head jamb for level. If the other side jamb needs to be picked up a bit, do so, and then nail it off. You may have to cut off uneven side jambs with a backsaw. This happens, for example, when you want side jambs to sit directly on hardwood flooring. If you are using carpet or vinyl, a small gap under a side jamb will be covered by the finish flooring material.

At this point, retrieve the bottom plate. This plate is the exact length needed to fit on top of the door jamb, where it now becomes drywall backing. It is fixed in place by toenailing through the backing into the trimmer. The backing eliminates the need for header cripples, provides backing for drywall nailing, and makes the head jamb extra secure.

The last step is to cross-sight the side jambs to ensure that they are parallel to each other. If cross-sighting is not done accurately, the door will hang improperly. Cross-sight by eye instead of using the time-consuming method of holding a string diagonally from corner to corner. Stand along the wall and sight along the side jambs to see if they line up with each other (drawing next page). If they don't line up, tap the out-of-line jamb into place by hammering on the bottom plate of the wall until both sides line up. If the building's frame has been properly plumbed and lined, very little correction will be needed to ensure that the jamb sides are in perfect cross-sight.

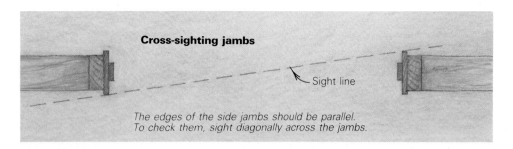

Cross-sighting jambs

Sight line

The edges of the side jambs should be parallel. To check them, sight diagonally across the jambs.

Once you have accurately cross-sighted the side jambs, the door jamb is completely set and ready for its door. The entire jamb-setting operation takes only five or six minutes.

The door hanger's bench—The door hanger's most prized possession is his bench, an ingenious work station that is easy to carry from one job to the next (photo left). While no two benches are exactly alike, they will share certain features. Note how the four corners are covered with carpet to protect the door when it is laid flat. Dowels slide through the legs of the bench to support doors of different widths while they recline on edge. A metal hook at one end of the bench keeps the door from falling sideways, and the corners of the bench are wrapped down the sides with carpet to protect the doors where they lean against the bench.

Inside the bench are bins that contain a router, circular saw, electric plane, electric screwdriver, ½-in. drill and lock jigs. Every tool needed to hang a door is close at hand. And on one side of the bench there are multiple electrical outlets to service all the power tools, eliminating time-consuming plugging and unplugging.

Hinge template—Once the drywall is on the walls, the first step in doorhanging is to rout out the hinge gains (mortises) on the jambs. Production door hangers like Royal Schieffer do this with a hinge template guide (bottom photo, facing page) and a router. One-piece hinge templates can be purchased (see *FHB* #31, pp. 28-31), but door hangers prefer to make their own. Before it can be used, the commercial template has to be secured to the side jamb or door edge with two small pins or nails. Royal's homemade template eliminates this step because it can easily be held in place on the jamb by foot, or on the door edge by hand. When routing hinge gains on the jamb, the template's metal tabs register against the door stop to align it in the horizontal position. The template is shoved against the head jamb, where a round-head screw driven into the top of the template's body acts as a spacer to hold the template ⅛ in. away from the jamb. The spacer gives the door proper clearance between itself and the head jamb.

To rout the hinge gains on the jamb, Royal simply holds the template in place with his foot and runs the router bit around in the hinge guide. Royal uses a ¾-hp router fitted with a collar that rides along the inside edges of the template. The two-flute ½-in. carbide bits that he prefers are made by Paso Robles Carbide (731 C Paso Robles St., Paso Robles, Calif. 93446). Remember that a router is a high-speed power tool and throws out wood chips. Always protect yourself against eye injuries.

Interior doors require two 3½-in. butt hinges. The top of the top hinge is 7 in. from the head jamb. The bottom of the bottom hinge is 11 in. from the bottom of the door. If

A door hanger's workbench contains all the tools necessary to do the various parts of the job, along with places to support the door while it's being worked (photo above). Here Royal Schieffer uses a ½-in. drill and a lockset jig to make short work of the holes needed for locksets and dead bolts. Just below the lockset jig are two registration marks drawn in pencil on the bench rail for locating the jig's positions. The hollow in the center of the bench allows one person to stand inside it and easily lift it from job to job.

To protect fragile door skins, Royal Schieffer's saw has a piece of plastic laminate contact-cemented to its base. The bent end of the metal flange to the left of the blade holds the wood fibers in place during the cut.

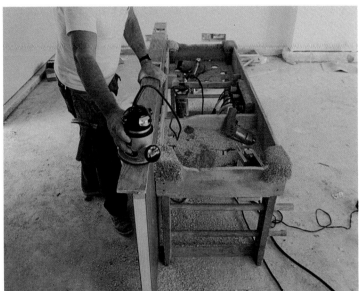

The same template used to rout the gains in the jambs is used to rout the gains in the edge of the door. The end of the template is flush with the end of the door. Note the template's round-head screw: it ensures 1/8-in. clearance between the door and the head jamb.

called for, a third hinge is centered between them. Heavy doors need three 4-in. butt hinges. Once the jambs are routed, it's time to fit the doors.

Fitting a door—Experienced carpenters usually don't need to check the blueprints to see which way a door swings. For example, a bedroom door most often swings in. Check the location of the electric switch. When the door is open, you must have easy access to this switch. If you have any doubts, check the plans and then make a mark on the floor indicating which way the door opens.

At this point set the door in place in the opening in preparation for scribing it to fit. The door is set on a block prior to scribing. Interior doors typically are lifted ½ in. off the floor if vinyl will be the finish flooring, ¾ in. for hardwood flooring, and 1½ in. for carpet. Many door hangers carry a 2x4 block that has been cut into ½-in., ¾-in. and 1½-in. steps for just this purpose.

Royal holds the door in place with a homemade door anchor placed over the top of the door and hooked on the inside of the jamb (drawing above). This tool isn't on the market, but it's easy to make and it's indispensible to a door hanger. It holds the door firmly in place, freeing your hands to complete the scribing process.

The easiest way to scribe a door is with a short, round pencil or a flat carpenter's pencil. Hold the pencil on the jamb and make a mark on both sides and on the head of the door. Unless your door has stiles that extend beyond the bottom rail, there is no need to scribe the lower end of the door. The door can be shortened by removing the excess from the top, requiring only one cut, not two. Mark the hinge location on both the jamb and the door and place the door flat on the bench.

Cross-cutting a door—Cutting across the vertical grain of a door has always been a touchy process. Done incorrectly a saw blade can break slivers loose from the door's veneer, resulting in a ragged cut. One way to avoid "tearout" is to lay a straightedge across the door and score the cutline with a sharp knife. This works quite well, but it's time-consuming. Another way to make the cut is with a saw fence we call a "shoot board." This is a straightedge with a fence screwed to it. The distance between the straightedge and the fence is exactly equal to the distance between the left side of the saw blade and the left edge of the saw's base. Clamp the straightedge to the cutline. It holds down the veneer as the fence guides the cut. A third method is to modify a saw with a homemade flange that is affixed to the top of the saw's base (photo above left). The flange is adjacent to the saw blade, and it rides lightly on the door during a cut, preventing the veneer from lifting. It really works.

Running a metal saw base over a door often damages or scratches the finished wood. A way to avoid such damage is to glue a

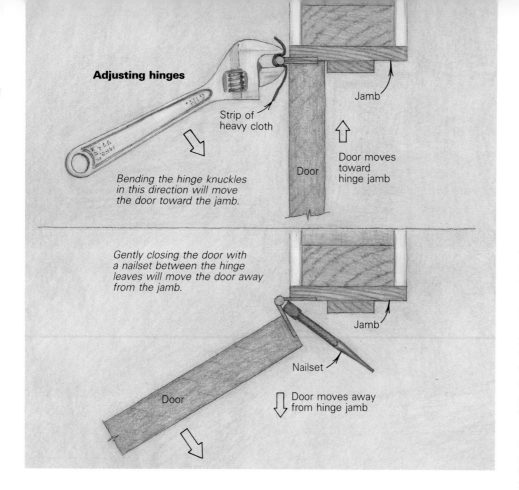

Adjusting hinges

Bending the hinge knuckles in this direction will move the door toward the jamb.

Strip of heavy cloth

Jamb

Door moves toward hinge jamb

Door

Gently closing the door with a nailset between the hinge leaves will move the door away from the jamb.

Jamb

Nailset

Door

Door moves away from hinge jamb

Hanging a door—Like many other carpentry techniques, door-hanging gets easier with practice. An experienced door hanger doesn't have to pull the pins and split the hinge leaves—even with heavy, solid-core doors. A typical installation takes about five minutes.

Take the door, along with your battery-powered screw gun, to the opening. Place the bottom of the door on the end of your foot so that you can insert the top hinge in the jamb gain. Once the door is in position, secure the top hinge into the jamb with one screw. Next, swing the door back parallel with the wall. Put your toe against the bottom edge, push the door plumb and, like magic, the other hinge should drop into its gain. Now it's just a matter of driving home the remaining screws.

Before doing anything else, check the door for fit. The door should fit almost perfectly, with little need for fine-tuning. Because the top and sides of the door are beveled, it's easy to shave the door a little with a block plane to make the clearance the same all the way around the door.

Sometimes the door will be the correct size, but the hinge keeps it too close to one side. Carpenters used to correct this by putting a piece of cardboard behind the hinge, but nowadays there's an easier way.

Over the years, door hangers have noticed that hinges vary from case to case. Hinges from one box may hold the door just a little closer to the jamb side than hinges from the next box. To adjust for this you can spring the hinges slightly. To move the door away from the jamb, stick the butt end of a nailset in the hinge and close the door gently (bottom left drawing). This opens the hinge a bit. To move the door closer to the jamb, use an adjustable wrench on the hinge knuckles (top left drawing). To keep from marring a hinge, use a piece of heavy cloth with the wrench. At times, I've heard criticism that this isn't craftsmanlike. Personally, I think the problem isn't with the craftsman, but with the variations in hinges. If your client is providing you with first-quality hinges, this problem shouldn't arise.

With the hinges adjusted, it's time to make one last check by cross-sighting the jambs. Other tradesmen can sometimes bump a wall and knock jamb sides out of parallel. The finished door, when closed, should be flush with the jamb on all three sides. If it's not it can usually be corrected by moving the wall a little. You can do this by placing a block against the drywall at the bottom plate and tapping the block until the door is flush with the jamb. To make sure that the wall stays in the desired position, drive a small wedge under the bottom plate. The weight of the building will keep everything in place. The door is now ready for a lock, stops and casing. □

piece of plastic laminate to the base of your door-hanging saw. It's also a good idea to attach laminate to the fence of the electric plane. Rubbing a little paraffin on the laminate makes these tools float over the work.

When you cut the top of the door, bevel it about 2° to the inside. Do this for two reasons. First, it eliminates problems that occur from paint buildup over the years. Second, once the door is hung, you may have to take a bit off the top in places to ensure proper clearance all the way across. It's easy to do this with a block plane without removing the door. All you have to do is take a bit off the high edge—it's unnecessary to plane the entire head. The bevel gives plenty of clearance on the stop side. Hangers refer to this process as "fine-tuning the door."

At the bench—Once the door has been crosscut, it's time to place it on edge at the bench. The doorhanger now fits and bevels each side with a power plane. Doorhangers typically set the fence of the plane at 3° to achieve the proper bevel. Everyone knows that it's necessary to bevel the lock side. But it's also important to bevel the hinge side to prevent difficulties with paint buildup. And what's more, if the jamb is cupped, you won't have problems with hinge bind.

The amount of clearance between the door and the jamb depends on the region in which you live. In most situations, leave a little less than a $\frac{1}{8}$-in. gap at the top and on the sides. You'll automatically do this by removing the scribe line with the plane set at the 3° bevel. If the house is in a humid climate where the door is going to absorb moisture, take off another $\frac{1}{16}$ in. to keep it from sticking when it expands. In dry climates, the clearance can be less.

As you cut to the scribe line, keep a thumb on the front lever of the plane. Watch ahead of the plane, raising or lowering the cutter to the required depth. If there is an excess of material to be removed, you may have to make two or three passes with the plane.

Now turn the plane so that its cutter is at a 45° angle to one of the door's edges and put a slight chamfer on the edge. Done properly, all the door's edges can be dressed with the plane to make them look like factory edges. A pass or two with sandpaper puts the finishing touch to this stage of door hanging.

Routing for hinges—The next step is to rout the hinge gains on the door and install the hinges. Hold the template flush with the top of the door and cut the hinge gain with the router (bottom photo, previous page). Contemporary hinges have rounded corners so there is no need to chisel out the corners of the gains. Place the assembled hinges in position and drive in the screws with an electric screwdriver. Pilot holes are unecessary unless you're installing a hardwood door. With a little practice, you should be able to install the hinge perfectly (all of this is done right at the bench).

When you're finished on the hinge side, flip the door over to the lock side. The position of the locks—36 in. from the floor for the door knob, 42 in. for the dead bolt—is marked on the bench. There's no need to measure each door individually. Simply register the jig to the marks, clamp it in place (photo, p. 54) and bore the holes for the cylinder and the latch.

Larry Haun lives in Los Angeles and is a member of local 409, where he teaches carpentry in the apprenticeship program.

Hanging an Exterior Door

From framing the rough opening to mortising for hinges, installing a door requires patience and precision

by Jared Emery

An exterior door has to fit its jamb tightly, though not so tightly that it causes a shoulder separation when you try to open it. You'll have few problems if you're using a prehung door—one that's already fitted to its jambs. But if your roughed-in opening is an odd size or you want to use your own special hinges, say, or you want jambs made out of something other than stock pine to match your casings or paneling, you may want to hang your own door. This can be an awkward and frustrating task, but getting the proper fit is largely a matter of cutting the door to a median dimension that will accommodate seasonal changes in the wood.

The rough opening for the exterior jamb should be framed in larger than the door by the total thickness of both legs of the jamb plus ½ in. on either side for shim space. For example, a 3-0/6-8 door with a jamb made from 1-in. stock would require a rough opening of 39 in. wide and about 82 in. high.

The legs and head of the jamb should be 1-in. thick. The sill should be of 6/4 stock. These units are available unassembled, prerouted with the sill planed and the stop in place. Or you can make your own.

First give the jambs a primer coat and then assemble the frame, using waterproof glue and three 12d casing nails through each side jamb into the head. To avoid splitting the wood, don't nail within about ½ in. of the jamb edges. Square up the frame and then nail the sill in place the same way. Finish by driving two 16d casing nails through each corner of the head and the sill into the jambs.

Next you have to prepare the rough opening

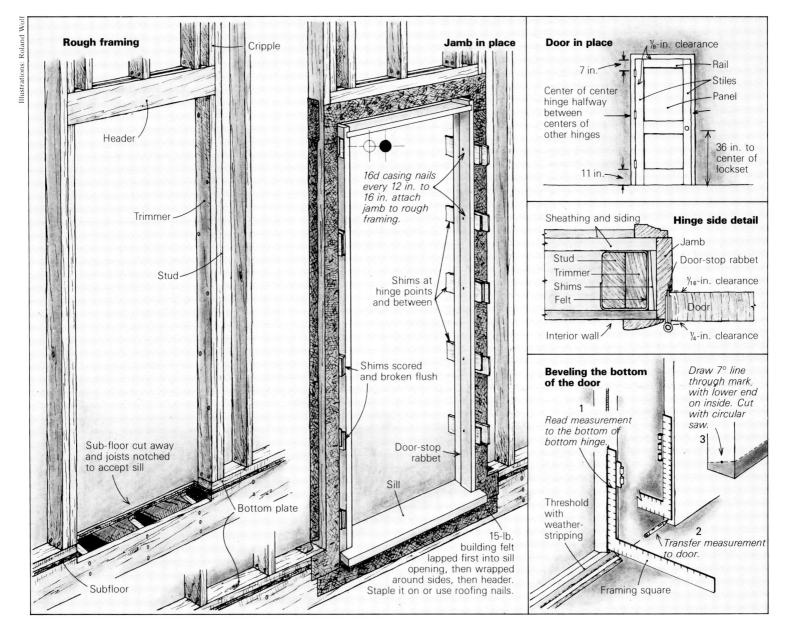

to receive the frame. Wrap the studs of the opening with 12-in. to 16-in. wide strips of 15-lb. roofing felt for extra insurance against water infiltration. The sill of the frame must finish out flush with the top of the finish floor. In most cases, you will have to cut out at least the subfloor, and you may have to notch or trim the rim joist as well.

Fitting the frame to the opening—The rest of the job depends on how well you set the frame. The two most common errors in this work are racking the frame in its opening, and plumbing one leg while leaving the head out of level. Once the frame is within the opening, level the head by shimming beneath the sill under the appropriate leg. It's best to use two wedges when you shim, driving one in from each side. Next, shim the hinge jamb plumb, setting double wedges at the hinge locations; drive 10d finishing nails only partway through the jamb into the trimmers to hold it snugly against the wedges. Plumb the other jamb the same way, add several more shims along each side to keep things steady, recheck for square, then drive and set 16d casing nails through the center of the jambs into the trimmers every 12 in. to 16 in. Then remove the 10d nails, which held things temporarily in place.

Up to this point, what you've done would apply equally to prehung doors or to those you would hang yourself. If you were working with a prehung door, you would already have had to determine which way you wanted your door to swing. If you're hanging your own, you could wait this long to decide whether the door will open on the right-hand or left-hand side. All exterior doors open to the inside. If there is a light switch on the wall, the door should swing away from it so someone entering can reach inside to turn on the lights.

One of the best kept secrets of carpentry is how to determine the "hand" of a door. This is important when you're ordering some lock hardware or a prehung unit. Simply stated, when the door is closed and you are standing on the inside, if the knob is on the right, it's a right-hand door.

Fitting the door to its jamb—This part of the job requires that you carefully measure the opening and know the tolerances demanded by the weatherstripping you intend to use. Of the many types available, the one I like most is the spring-type weatherstripping. Made of bronze-colored light-gauge aluminum, it has a nailing surface and a flange that angles back toward the outside, against which the door will close. If you use this material, you should size the door $\frac{1}{4}$ in. less than the jamb width. This dimension allows $\frac{1}{8}$-in. gaps at both the hinge and lock sides. The optimim clearance for the lock stile is $\frac{3}{16}$ in., but it's a good idea to leave the door a little wide at first, so that any irregularities in the jamb may be planed into the door after it has been hung.

At this point, you should do any trimming on the hinge side. This will leave a full stile for mortising the lock hardware. This is especially important if the door has a glass panel in it. Having a lock mortise break through the stile into a glass panel is guaranteed to be a heart-stopping experience.

Finally, plane a 5° bevel on the lock stile, so that the leading edge of the door will clear the angled portion of the weatherstripping. I use an electric Rockwell Portaplane; a well-sharpened jack plane works well, too.

The height of the door should be $\frac{1}{4}$ in. less than the total jamb height, for $\frac{1}{8}$-in. clearance at the head and just enough clearance at the bottom to allow it to close. (The bottom of the door should be cut to fit its weatherstripped threshold later, after you hang the door.) Check the jamb again for squareness and plumb before you do any cutting.

After the door has been fit to the jamb, it is ready to be hung. Hinges should be selected according to the size of the door. Most exterior doors are solid core and $1\frac{3}{4}$ in. thick. Building codes require them to be at least 3 ft. wide. The best hinges to use to support this much weight are 4-in. by 4-in. loose-pin butt hinges.

Installing the hinges—Place the top of the top hinge 7 in. from the top of the door, the bottom of the bottom hinge 11 in. from the bottom of the door, and a third one centered between the two. If you're using a $\frac{3}{4}$-in. thick casing on the inside of the jamb, you should leave $\frac{1}{4}$ in. of the hinge leaf outside the mortise. This brings the outside of the hinge's barrel into alignment with the outer edge of the casing, and allows the door to swing 180° without running afoul of the casing and levering the hinges loose.

For laying out and mortising hinges, the door must be held securely with the hinge side up. A door buck is best for this (see the box below) though you can brace the door with a sawhorse by nailing a strip of wood to each. First mark out the hinge positions and then cut the mortises, as explained on the next page.

Next, mount the hinge leaves. Drill the pilot holes for the screws just slightly off center, toward the closed edge of the mortise. The wedging action of the screw head will pull the hinge tightly into the mortise.

When the hinge leaves are in place, the door can be set into its jamb, wedged up to the correct height and shimmed against the hinge side of the jamb. Using a utility knife, score the jamb slightly above and below each hinge. The knife-marks indicate the exact location of the hinge, something a pencil line cannot do.

It is good practice to make the mortises on the jamb $\frac{1}{16}$ in. or so narrower than the mortises on the door. This is especially true if either the door or the jamb is to be painted. This difference in widths allows for the thickness of the paint between the door and the stop and for the expansion of the stop. Mortise the jamb the same way you mortised the door stile. But you'll find it more awkward to do precise work on a vertical surface than on a horizontal one.

Always put the pin in the top hinge first, so the door can hang while you catch your breath and ease your cramped arms and aching hands. If the mortises have been cut accurately, the hinge leaves will line up and fit together. If one of them doesn't, though, loosen the screws on both leaves of that hinge and tap them together with a hammer. Then insert the pin and retighten the screws.

There should now be $\frac{1}{8}$-in. clearances on both the hinge and lock sides of the door. But

---

There are two common ways to hold a door while you work on one of its edges. Door bucks work well for me. You can buy these, but I make mine out of triangles of 2x8 stock, usually stair-stringer scraps. I nail two of these a door-width apart on a platform of $\frac{1}{2}$-in. plywood long enough to hold the door. Then I fasten 2x4 blocks to each corner and in the middle as legs. The weight of the door deflects the plywood, pressing the tops of the triangles against the door. This holds it steady, and you can take it out with no fumbling around.

The other method is to set up the door parallel to a sawhorse and tack a small strip of wood to both the horse and the door with 4d nails. A single strip will hold the door, but this method is a bit more cumbersome than using a door buck, and the nailed strip gets in the way if you're planing. —*J.E.*

Building a door buck

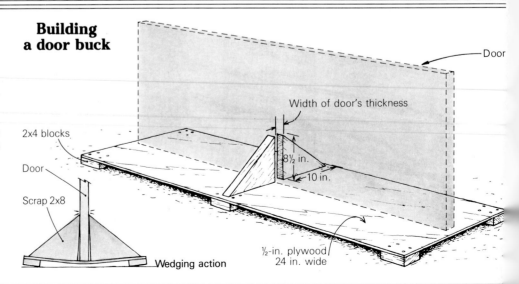

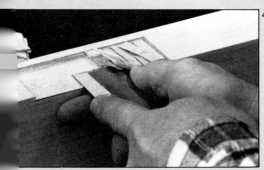

Mortising butt hinges

For mortising hinges, many tradesmen like to use a router. But with a good butt chisel, a hammer and a little practice, you can cut mortises for hinges on a single door as fast as you can with a router. Router setups take time, and you'll probably have to trim the cut with a chisel anyway. So why not do the whole job with a few hand tools?

Lay out the mortise first. The two lines across the grain must be struck with a knife to cut the wood fibers cleanly. The two lines with the grain—one for the back edge of the leaf, the other for the depth of the mortise—should be made with a scratch awl or other pointed tool. Don't use a pencil only, or you'll get imprecise results, and risk splintering the wood outside the area you want to excavate.

Determine the depth of the mortise by figuring how wide a gap you need between the door and the hinge jamb. For an interior door with no weatherstripping, you may want to cut the mortise slightly deeper than the leaf is thick, to make the gap neat and minimal. For an exterior door, the mortise will be as deep as the leaf is thick. The barrel of the hinge should be set out from the door enough so that when the door opens, it clears the casing.

With the layout done, score the wood in about 1/4-in. wide increments down the length of the layout. Use a sharp chisel as wide as or wider than the mortise, and hold it at about 45°, as shown in photo 1. With a single blow from hammer or mallet, cut the wood tissue to within a hair of the depth line. About 10 of these cuts should do for a 4-in. butt. The closer you hold your chisel to 90°, the more force required to sever the wood fibers. Do not at this stage try to cut straight down on the lines at the ends of the mortise.

Next hold the chisel vertically, so its bevel is away from the line, and cut the back wall of the mortise with a few light taps of the hammer (photo 2). If you try to make this cut before scoring up the wood inside the mortise, you will probably split out the wood along the line, and ruin the whole job. This will happen because the still-intact waste will not yield and will force the chisel to act like a wedge instead of a cutting tool.

Having cut the back wall of the mortise, you should now pare away the scored tissue. Begin about 1/2 in. away from one side, index the edge of the chisel (bevel up) in the depth line, and pare straight back to the rear wall (photo 3). This should not require much force, which means that you shouldn't risk running the chisel past the rear wall of the mortise. So be sure your chisel is sharp, and that you've scored the wood to the right depth before you begin horizontal paring.

The first cut should give you a nice flat surface the full width of your chisel. As shown in photo 4, you'll use this surface to register the chisel for subsequent cuts to get the whole mortise in the same plane. When you've pared away most of the waste, chisel the end walls square to the bottom. Again, don't try to make these vertical cuts until you've gotten the adjacent tissue out of the way.

When all's done, the hinge should fit snugly (photo 5). Another advantage in mortising with a chisel is that you can quickly make slight adjustments in the depth of the mortise, something that would take a lot of fiddling if you used a router. —*John Lively*

the optimum clearance on the lock side is 3/16 in. Some carpenters like to trim the door to this tolerance before hanging it, but I've noticed that the general tendency is to trim off just a little too much. (You can shim out the hinges with cardboard in this situation, but that's a step I'd rather not have to take.) There are often slight irregularities in jambs that have to be accommodated, too. I prefer to check the door once it's hanging, scribe for the 3/16-in. gap, then take it down again and plane to the line, maintaining the 5° bevel.

Hardware and weatherstripping—Next, let in the lock hardware and apply the weatherstripping. It is a good idea, when mortising for the lockset, to cut the mortise slightly deeper than it has to be. This will give you some room for adjustment if the door needs any planing in later years.

A high-quality lockset comes with an adjustable faceplate that can be set to the bevel on the door. Less expensive locksets don't, and setting the lock deeper into the door keeps the faceplates of moderately priced locksets from protruding at the lower side of the bevel.

Nail weatherstripping to the jamb with the flange angled toward the outside every 2 in., using the small-headed brads supplied. The spring flanges of bronze weatherstripping have to be mitered at the corners to fit, and it must be cut around the strike plate on the lock side of the jamb. I nail the flanges down above and below the strike to prevent excessive wear.

There are several satisfactory types of weatherstripping available for the threshold. The most common is an aluminum saddle with a rubber or vinyl insert. The saddle is notched to fit around the door stop, casing and jamb, so that the rubber insert will lie beneath the door. Once the saddle has been fitted and secured to the sill, the door must be cut to fit.

To find the cut-line across the bottom of the door, set a framing square against the rubber insert with the tongue of the square against the hinge-jamb leg. This will tell you if the saddle is lying square to the jamb, and will also give you a measurement from the rubber insert to the bottom of the bottom hinge.

With the door once again lying on the sawhorses, transfer this measurement to the center of the edge of the hinge stile. Then draw a 7° line through this mark, the lower end of the line being on the inside of the door. Where this line comes out on the inside, draw another line across the bottom face of the door. This is the cut-line. If the door has a veneer face, score it first with a knife to prevent chipping. Then set a circular saw at 7° and cut the door bravely. If you're going to paint the door, prime the cut edge at once. If you're going to stain it, do so at once and then apply varnish or one of the urethane finishes. Don't let a new door hang for long without a sealer or prime coat.

Rehang the door and check its fit. It should close snugly and give slight resistance when it is opened. □

Jared Emery is a writer disguised as a carpenter. He lives in Charlottesville, Va.

Ordering and Installing Prehung Doors
Precise results require careful planning

by Steve Kearns

Our crews consider installing anything but a prehung door on our jobs to be a throwback to the Dark Ages. With a few tricks and some basic carpentry skills, you can install a prehung door in half an hour or less. Sure, you pay more to have your doors prehung—our door shop reports the average price is $5 to $7 per door—but we find it unimaginable that anyone could save money or have better quality control by doing the hanging on site.

The walk-through—My first piece of advice is this: Don't order your doors off the plans. Details like door size, handing (hinges on the right or the left side) and the direction of swing can change as the walls go up. So we wait until all the door openings are roughed in before ordering the doors. There is an exception to this rule: We have our shop order the blanks for any custom doors that have a long lead time as soon as possible. The handing and the jamb widths may change, but the doors probably won't.

For interior doors, we frame our openings 2 in. wider and 2 in. taller than the door size. For example, a 2 ft. 6 in. by 6 ft. 8 in. door requires a 2 ft. 8 in. by 6 ft. 10 in. rough opening. The gaps allow for ¾-in. thick jambs and ¼ in. of shim space on each side. An exterior door comes with a threshold (photo above) and typically needs an additional 3 in. of height to allow for its thickness. Once the house is framed and the change orders accommodated, you know for certain the size of every door and the thickness of the walls.

Now is the time to take a walk, plans in hand, with your door-shop representative to lay out and specify every door in the house. Our local shop, Pozzi of Idaho, has an order form for this purpose. The door specifications include its size, style, handing, jamb species and width, hinge size and finish, threshold (if needed), sweep, lockset and dead-bolt bores, type of hardware

Standing by. Whether an interior or an exterior door, the pertinent information is written on the outside of the jamb. The notes for this exterior door include the contractor's name, the width of the jamb, the handing (L.H.), a code for the hinge type and the width and species of the threshold material. Three lines of Geocel caulk provide the weatherproof seal under the threshold.

and any special details, such as the need for paint-grade or stain-grade jamb stock. If the form is filled out completely and accurately, you should be able to eliminate one bottle of aspirin from your list of supplies for the job. Making a mistake on the order form usually means you'll have to reorder or remodel something, then explain why to a long-faced client who showed up 10 minutes after the doors arrived.

Back to our walk-through. Our door man carries his order form, a pencil with an eraser, a carpenter's keel and a tape measure. I carry the blueprints. A good set of prints will have the doors numbered on the floor-plan pages and listed in a door "schedule." A comprehensive door schedule will include most of the information listed in the order form, but if you're trusting enough to order straight from the door schedule, keep the aspirin.

We start at the opening for door #1, measuring to check the called-out door against the framed opening. Assuming there's no difference, we mark the door's number and indicate its handing on the trimmer by drawing a rectangle with an *X* through it in the position of the top hinge. At the same time, our door man notes the rest of the pertinent specs for this door.

We move throughout the house in this manner until we have laid out every door. There should be agreement between the order form, the number of doors shown on the plan and the number listed in the schedule. If these numbers disagree, we reconcile the discrepancies.

Taking delivery—Now the day comes when the doors arrive on site. The scheduling is perfect: The drywall is hung, taped and textured; the walls are painted (we put masking tape over all the keel notes on the rough openings before the painter sprays the walls; otherwise, they will disappear); the floor is swept; and carpenters are waiting to take the doors off the truck. The shop has written each door's number on the outside of its jamb, and we take each door to its designated opening. We usually have someone reading the plans to direct traffic as the doors are unloaded. If the scheduling is off, we store the doors out of the way and under tarps until all the prep work is complete.

Installation—Once the doors are distributed, a carpenter begins installing each one by checking the basics: The floor should be level, the header

1. Flushing up. Carpenter Alex Harakay nails the strike-side jamb while checking that the exterior siding and the edge of the jamb remain flush.

2. Missing the shims. Pairs of nails are driven below the shims rather than through them, so the jamb can still be adjusted.

3. Fine-tuning the reveal. With the door closed, Harakay adjusts the shims to achieve an even space between the door and the jamb.

4. Bullnose corners. For a rounded corner detail, metal corner bead is tucked into factory-cut kerfs in the edges of the jambs.

square with the trimmers and the trimmers plumb in both directions.

If the floor is not level, you'll need to know which jamb leg needs to be adjusted to correct for it. Interior doors typically come with a 1¼-in. gap between the bottom of the door and the bottom of the jambs. That gap allows for a carpet pad and a carpet. If the flooring isn't as thick as a carpet, cut the bottoms of the jambs so that the door will clear the finished floor by ¼ in. to ⅜ in.

The hinge jamb is the first to be affixed to the framing. Ideally, the trimmer on the hinge side of the door should be straight and plumb, allowing the hinge jamb to be nailed to it without shims. Each jamb is affixed to its trimmer with three pairs of 16d finish nails. If the trimmers are slightly out of plumb in the plane of the wall, shimming the door jambs will be required.

If the opening is just a little out of plumb in the plane perpendicular to the wall, our rule is "go with the wall." That way the door casings will fit flush with both the wall and the jamb. If the trimmers are out of plumb in different directions perpendicular to the wall, we fix the wall before installing the door.

When a prehung door leaves our supplier, it has a couple of little shipping blocks under its jambs to protect the corners, as well as thin spacers between the jambs and the edges of the door, and several 6d duplex nails driven through the jamb into the door edges to keep everything aligned. The nails and the blocks have to be removed before the door is slid into its opening. If the hinge screws poke through the jamb, we cut them flush with a reciprocating saw so that the jamb can be tight to the trimmer.

Let's say our rough opening isn't perfect. Because we checked the hinge-side trimmer before we inserted the door, we know which way it's out of plumb and whether we should start nailing at the top or the bottom. In this example, the floor is level and we will need a shim near the bottom to plumb the jamb, so we start by nailing at the top of the jamb, just below the hinge. The door has to be open while the nails are driven home. We support the door in the open position with a stack of shims to keep it from pulling the jamb out of alignment.

A jamb for a typical 2x4 wall is 4 9/16 in. wide. That dimension equals the 3½ in. of the stud and two layers of ½-in. drywall plus 1/16 in. The extra width makes it easier to get the casings to lie flush on the jamb edges with no gaps. The jambs should be checked periodically for alignment with the walls as the nails are driven through the jambs into the trimmers.

Next, from each side we insert a pair of shims just below the bottom hinge and plumb the jamb with a 6-ft. 6-in. level. We "over-plumb" it slightly because nailing moves the trimmer away from the door opening a little. The nails should pass just below the shims, allowing the shims free movement for adjusting the jamb in and out. With the top and bottom of the jamb secured, we put our level on the jamb and shim the middle until it's flush with the level's edge. After setting the nails, we give the door another check for plumb. If necessary, a little tapping on the shims should get it perfect.

To secure the strike side, insert pairs of shims snugly so that they'll stay put at the top and bottom (top photos, p. 61). Secure the shims with nails driven just below them. Now close the door and adjust the shims in and out to get the correct reveal between the jamb and the door edge (bottom left photo, p. 61). We match the thickness of the spacer blocks at the top of the door.

The door is now plumb and level, and its jambs are flush with the drywall on each side. A key step (and one that is often left out) is to replace at least two of the top hinge screws and one from both the middle and bottom pairs with screws that are long enough to bite at least 1 in. into the trimmer. The trimmer, not the jamb, should hold up the door.

Fitting the stops—All that remains is setting the door stops, which come from the shop loosely attached to the jambs. For this task we'll need an example of the strike plate and the lockset that will be installed in the finished doors. We screw the strike plate into the precut mortise in the jamb and insert the latch bolt into its mortise in the edge of the door. The rest of the lockset is not necessary for this fitting. Now we close the door tightly to the front of the strike plate while nailing the stops in place with a finish nailer. If the doors and the jambs are to be painted rather than stained and lacquered, we insert a matchbook cover as a temporary spacer between the door and the stops to allow for the paint buildup.

This last step of temporarily installing the hardware may seem like overkill, but we have found it the only way to ensure that the finished doors will close firmly against the strike plate and the stops. The door is now ready for the casings to be installed by the trim carpenters.

The exception to the sequence I've just explained is the bullnose casing detail. A lot of the houses we've built lately have radiused corner beads, which require the jambs to be installed before the drywall. The corner bead fits into kerfed jambs (bottom right photo, p. 61). The door shop takes care of the kerfs, and we have them take off ¼ in. per side from the width of the jamb to allow the metal to wrap smoothly from the drywall to the jamb. For example, the jamb for the 2x6 wall shown here is 6 1/16 in. wide.

Finishing the doors—Our painters usually take the doors off to finish them. And we have them number each door—inside the lockset bore is a good spot. They remove the hinges and the screws from both the door and the jamb and put the hardware in bags for safekeeping.

It takes longer to describe this process than to do it. When I do my estimate, I allocate an hour for installing each prehung door, including unloading and stocking them. My carpenters almost always beat the estimate.

Finally, store your painted doors until the carpet is laid. Now you're ready to go around in your socks and put each door back in place and install the locksets and strikes—all of which will fit perfectly because you ordered it so. □

Steve Kearns is a contractor based in Ketchum, Ida. Photos by Charles Miller.

Plumb-Bob Door Hanging

When the level won't work

By Scott Wynn

Few topics elicit more heated discussions among carpenters than the "right" way to hang doors. Now I'm going to throw another technique into the fray: hanging doors with a plumb bob.

On one of my early jobs as a carpenter, I had to hang about 25 doors in a remodeled Victorian. The new framing was poorly done. Many of the openings varied in width, and some tapered so much that there was zero clearance for shimming. The trimmers weren't even close to being plumb (in either direction). To make matters worse, as I nailed the jambs of the prehung doors to their lousy trimmers, the jambs began to twist, and they took on high spots between the shims that threw off my level readings. I'm sure there are ways to compensate for this, but I was having a hard time inventing them.

Then my boss came around. Perturbed at the slow progress, he asked, "Why don't you use a plumb bob?" That was the way he'd learned how to hang a door, and he was a little surprised that I wasn't using one already. Here, with some things I've learned since, is what he showed me.

Begin the job by removing the door from its hinges. Set the jambs in the opening and leave the stretcher on if you can. Check the floor for level. If it's out, temporarily shim the low side of the jamb until the head jamb is level. Transfer the distance you raised the jamb to the bottom of the other jamb and mark it. The amount you cut off the bottom of the jambs will depend on the finish flooring. I figure 1 in. for carpet and ¾ in. for hardwood.

Pull the frame out of the opening and trim the bottoms of the jambs as necessary. Remove the stretcher and the door stops and put the frame back in the opening. Now take a pair of shims (always try to use them in opposing pairs) and place them above the top hinge. In the portion of the jamb that will be covered by the stop, drive a 16d casing nail just below the shim—not through it—into the trimmer (drawings right).

Before you go any further, take a look at how the opposite jamb sits in relation to the corresponding wall. If the jamb is out of alignment, the cripple is probably twisted, and you will need to put in an odd shim to throw it back (drawing below right). A twisted trimmer usually requires an extra shim at each shim point.

Now tie the plumb-bob string to the nail so that the bob clears the jamb and falls just shy of the floor. I use a small plumb bob with about 8½ ft. of string tied to a short length of ½-in. dowel. I wind the unused line on the dowel and hang it out of the way over the ends of the top shims.

Place another pair of shims below the bottom hinges and secure them with a 16d nail driven below the shims and toward the hinge side of the string (to avoid hitting the plumb line). Shim and nail the

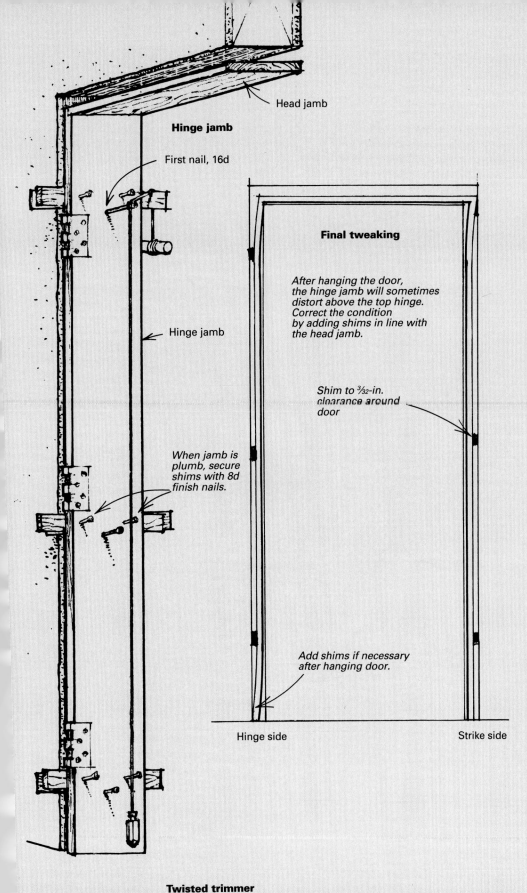

center hinge as well. You may have to add other pairs of shims, depending on the weight of the door, the number of hinges and the straightness of the jamb.

Now that the jamb is secured in several spots, note the distance from the string to the jamb where it is tied off (drawing below). It should be the same the entire length of the jamb, so drive the shims in or out to correct the distance to the string.

Once you've fine-tuned the shims to get the jamb perfectly plumb, drive in the 16d nails a bit more to make everything tight. Check everything again. When you are satisfied that the jamb is plumb, secure the shims with 8d (or larger) finishing nails. Now remove the string and drive in the 16d nails.

Rehang the door and shim the strike-side jamb and the head to an equal space around the door—usually $3/32$ in. I like to use the same technique of placing a single 16d nail below the shims and then setting them with a couple of finishing nails. Put the top set of shims on the strike-side jamb near the head (drawing left). This placement will help counteract the weight of the door pulling the head jamb away from the hinge side of the door. Likewise, the jamb sometimes warps below the shims under the bottom hinge. Add shims near the floor to straighten it out.

Remind yourself not to set any shims or nails (including casing nails) around the strike-plate area. Trying to work around a protruding nail when mortising for a strike plate will slow you down considerably.

After installing the lockset and strike plate, reinstall the stops. Start with the strike-side stop. Set it tight to the door at the head and bottom but flex it away from the door a hair shy of $1/16$ in. in the middle. This will allow the door to shut and latch crisply without it rebounding off the stop.

The hinge-side stop should be set about $1/32$ in. away from the door so that the door edge won't bind on the stop as the door is closed. Allow an additional $1/32$-in. minimum clearance for paint or varnish.

Now that I'm acquainted with its many uses, I always carry a plumb bob in my apron. It's easier to carry than a big level and a lot harder to knock over.

— *Scott Wynn is an architect/contractor who also designs and builds furniture in San Francisco, Calif. Drawings by the author.*

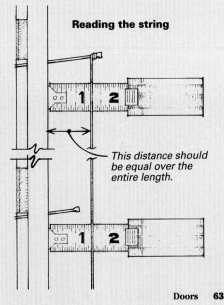

Installing Prehung Doors

An accurate level and a bucketful of shims will correct just about any out-of-plumb condition

by Jim Britton

CHECK THE ROUGH OPENING FIRST

O f all the tasks a trim carpenter faces, few offer the opportunity to transform the look of a house quickly from ragged edges to finished surfaces like installing prehung doors. It's the trim carpenter's version of instant gratification because once in the groove, a good trim carpenter can install a door, its jamb and all the casings in about 15 minutes. That's money in the bank for a pro, and a satisfying slice of sweat equity for the owner/builder.

But doors that squeak, bind, stay open or swing open by themselves are constant reminders of the fallibility of the trim carpenter. In this article I'll describe the methods I've settled on after 20 years in the trades for efficiently installing a typical prehung door and avoiding common glitches that bedevil a door installation. Like most homebuilding jobs, installing a door begins with checking work done before you got there.

Check the rough opening first—In a perfect world of accurate levels, conscientious framing crews and straight lumber, all rough openings are square, plumb and correctly sized. Because these three conditions rarely coincide, it falls to the trim carpenter to compensate for less-than-perfect rough openings.

Although there are exceptions, the rough opening should be 2 in. wider and 2½ in. taller than the door. Thus, the correct rough opening for a 2-ft. 6-in. door would be 32 in. by 82½ in. The extra space allows room for the door jambs and a little wiggle room to accommodate rough openings that are out of plumb. In my experience, rough openings are the same for both interior and exterior doors that are made by door manufacturers. Doors made by window manufacturers, on the other hand, sometimes require a different rough opening. If in doubt, check with the manufacturer before the framers start work.

Before installing a door, I inspect the rough opening to familiarize myself with its condition. First I check the dimensions to see if they are workable. Then I use a 6-ft. level to check that the two trimmers (the studs that frame the rough opening) are plumb (photos 1 and 2) in both directions. Sometimes the trimmers will actually be plumb in both directions, in which case the door jambs will be flush to the wall, and the casings will be easy to install.

But in some situations, the wall will be out of plumb in section, with the trimmers plumb in elevation. In this case, the door jambs will have to protrude slightly beyond the plane of the wall at the top and bottom on opposite sides.

Another common condition is the parallelogram-shaped rough opening. The wall may be plumb in section, but the elevation view of the rough opening is out of plumb. The net door width is usually ½ in. narrower than the rough opening. Therefore, I can install a plumb door in a rough opening that is up to ½ in. out of plumb. The jambs will fit snugly to the diagonally opposite corners of the rough opening. If the rough opening is more than ½ in. out of plumb, I use a sledgehammer to pound the trimmers into line. I can usually move a trimmer up to ½ in. without adversely affecting the drywall. This operation requires cutting back the sole plate once the trimmer has been adjusted.

The scissor condition, in which the trimmers are out of plumb in opposite directions in sec-

1 and 2. Read the rough opening. Before installing the door and its jamb, the author checks the trimmers on both sides of the rough opening with a 6-ft. level to see if the trimmers are plumb (photo facing page). If the jamb needs to project beyond the plane of the wall in order for the door to hang plumb, he notes the direction of the adjustment on the trimmer. To avoid mistakes, he marks on the floor the direction the door will swing.

3. Check the trimmers for twist. If the trimmer isn't square to the header, the door jamb will also be askew. Use a square to gauge the accuracy of the door frame.

Doors 65

tion, requires a more involved solution. Let's say one trimmer is ½ in. out top to bottom in one direction, and the other trimmer is out ½ in. in the other direction. This situation amounts to 1 in. of scissor. This condition is remedied by holding the jamb out ¼ in. at the top and in ¼ in. at the bottom on opposite sides of the wall. Do the opposite for the other trimmer.

The other condition I look for is twist (photo 3). If the trimmers aren't square to the header, the jamb will likewise be twisted. This condition results in hinge binding or a poor visual relationship between the door and its jamb after the installation. I take the twist out when I affix shims to the trimmer.

Next, put up the hinge shims—To begin an installation, I measure the height of the top and bottom hinges of the door from the bottom of the hinge jamb. My 6-ft. level makes a convenient stick to note the middle of the hinge positions (photo 4). These locations mark where I fasten my shims to the framing before the door goes in.

I shim the bottom location first with an appropriate combination of shims to bring the level plumb and to compensate for any twist in the trimmer. Then I move to the top hinge position, holding the shims in place with my level as I affix the shims to the trimmer (photo 5). Prenailed shimming makes handling the door easy and ensures that the door will automatically be plumb

in the elevation view. I use a 15-ga. or 16-ga. pneumatic nailer loaded with 1¾-in. to 2-in. nails for installing prehung doors. If you don't have one of these wonderful timesaving tools, use 8d finish nails instead.

Now it's time to squeeze the door and jamb into the opening. Remove any nails or straps used as bracing, and place the hinge jamb atop the thick end of a shim resting on the floor next to the trimmer (photo 6). Raising the jamb has three benefits: It eliminates squeaks by separating floor and jamb; it eliminates the problem of an out-of-level floor preventing the strike jamb from not coming down far enough to engage the lockset latch; and it eliminates (or minimizes)

NEXT, PUT UP THE HINGE SHIMS

4. Note the hinge positions. Using a 6-ft. level as a story pole, the author marks the centers of the top and bottom hinges.

7. Mark the jamb alignment on the shims. If you need to adjust the edge of the jamb in or out of the plane of the wall to get the door to hang plumb, make a note of the correct position of the jamb's edge on the top and bottom shims.

5. Affix the shims. Shims behind the top and bottom hinges make backing for the jamb. A single shim at the top compensates for twist.

8. Nail the jamb. Secure the jamb to the trimmer with a couple of nails right next to the hinges. The nails *must* pass through the shims.

the need to remove some of the door's bottom to accommodate finish flooring.

Once the jamb is in the rough opening, I swing open the door. If it's a troublesome installation, I'll block the door with a couple of shims. But typically I leave the open door unsupported. If the wall is plumb at the rough opening, I bring the edge of the hinge jamb flush with the drywall and nail it to the trimmer through the shims. If the wall isn't plumb, I compensate for the error by moving the jamb out equal amounts at the bottom and then the opposite direction at the top. I make pencil marks on the shims to note the correct alignment for the edge of the jamb (photo 7).

I affix the jamb in the correct position relative to the wall with a couple of nails through the top and bottom shims. Then, while the door is still open, I drive a couple of nails through the jamb right next to the hinges (photo 8). The hinge jamb and door now should be hanging plumb because they are held fast against the shims. If there is a middle hinge, shim it at this time, taking care not to make any changes to the already perfect alignment.

Nails are enough to keep the jamb of a hollow-core door from pulling away from the trimmer. But if I'm hanging a solid-core door, I run a 2½-in. screw through the jamb and into the trimmer next to each hinge. Because this step leaves a hole that no painter will be pleased to discover, I put the screws under the doorstops. Working from the bottom, I carefully pry away the stops (photo 9) and set them aside.

I sometimes run a 2½-in. screw through a hinge and into the trimmer. But I don't do this to keep the door from sagging: A properly hung door doesn't sag. Instead, I use the longer screw to straighten a warped jamb or to compensate for a hinge mortise that might be shallow. The longer screw will give me about $\frac{1}{16}$ in. of adjustment.

Secure the head jamb and strike jamb—Now that the hinge jamb is firmly secured to its trimmer, I close the door. Next I set the head jamb

SECURE THE JAMBS

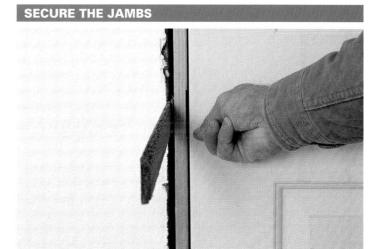

10. Equalize the strike-jamb reveal. Use shims placed next to the door latch to adjust the strike jamb in or out until the gap is consistent from top to bottom.

11. Reinforce the strike plate. After pulling the doorstop, run a screw through the strike jamb next to the door latch.

6. Don't forget the shim on the floor. Elevate the hinge jamb by placing it atop the butt end of a shim shingle as the door is lifted into position.

9. Remove the stops. Pry the doorstops from the bottom up. Then locate the screws that secure the jamb to the trimmer under the stop.

Doors 67

parallel to the top of the door by raising or lowering the strike jamb. At this stage of the game, a single, unshimmed nail through the strike jamb into its trimmer or through the head jamb into the header can help hold the parts in alignment while I assemble the correct combination of shims. A jamb held by a single nail still can be pried in or out as needed. A shim under the strike jamb also can be helpful.

When I've got the head jamb parallel with the top of the door, I check the reveal along the edge of the door and the strike jamb. I put a couple of shims between the jamb and the trimmer 6 in. down from the head jamb and adjust the shims until the gap, or reveal, between the door and the jamb is the same at the top corner. Then I shim the bottom of the jamb, 6 in. from the floor, and the center of the jamb opposite the strike plate (photo 10). Some door jambs are straighter than others. If I've got one with some dips and wows in it, I add shims as necessary to keep the reveal consistent. I add extra support to the jamb where the strike engages it. To do so, I pry away the doorstop and drive a 2½-in. screw into the trimmer (photo 11).

Replace the doorstop—The door is now where I want it and fully supported. With the door closed flush with the head jamb, I position the head stop on the hinge-jamb side with the help of a dime (photo 12). This ¹⁄₁₆-in. gap between the door and the stop helps to keep the door from binding on the stop and allows for paint buildup. I continue this space down the hinge jamb with the hinge stop, attaching the stop with my 16-ga. nails, 16 in. o. c.

I install the strike stop so that it just touches the entire height of the door with the door's face and the jamb flush over the full height. This system works well for a strike that has an adjusting tang. However, if the strike will be the T-mortise type, it will have to be installed first and the stop set to it.

Apply the casings last—I start casing a door at the top (photo 13) with the head casing set back ³⁄₁₆ in. from the edge of the jamb. The casing has 45° miters on each end, and the short side of the casing is ³⁄₈ in. longer than the jamb opening.

REPLACE THE DOORSTOP

12. Don't forget the stop gap. Use a dime between the closed door and the doorstop to gauge a consistent gap between the door and the stop. The gap allows for paint buildup.

APPLY THE CASINGS LAST

13. Casing starts at the top. Britton begins trimming a door by installing the head casing first. He affixes the casing with pairs of nails on 16-in. centers. One nail goes into the jamb, and another into the header.

14. This gap won't make the cut. When the door jamb and the wall are in slightly different planes, the casings don't lie completely flat. The tapered gap at the inside corner is the result.

That gives me the 3/16-in. reveal along both of the side jambs.

The tricky part of casing a door is dealing with the differential between the plane of the wall and the plane of the jambs when you've made allowances for an out-of-plumb rough opening. For example, this door had a head casing whose edge was recessed a bit from the plane of the wall. When I test fit the side casing, I came up with a gap at the inside corner (photo 14). To fix it, I undercut the miter with a disc sander (photo 15). This cut isn't a back bevel, however. In this case I removed material from the casing's face. Once I'd shaved the miter, I had an acceptable joint for paint-grade trim work (photo 16). To keep the adjoining casings in the same plane at the outside corner, I put a thin shim under them (photos 17 and 18).

I attach the side casings with pairs of nails, one into the jamb and one into the trimmer a couple of inches away. This nailing pattern helps ensure that the casing will lie flat. I nail the casings next to the hinges and the door strike because these spots are well-backed by shims. Nailing the casing at these points also reinforces the jamb.

As you can imagine, drywall edges can be a pain in the neck when the door jamb is below the plane of the wall. The hollow milled into the back of the casing is there to compensate for this situation. If the hollow isn't enough to accommodate the drywall, I use my hammer to "tenderize" protruding drywall edges.

If the floor is to be covered with carpet, I hold the side casings 3/8 in. above the floor. That gap gives the carpet guy some room to tuck the edges of the rug. It's a good idea to put a shim between the jamb and the trimmer at the bottom of the jamb if the room is to be carpeted. The shim keeps the jamb from being deflected by the carpet-layer's bump hammer as he tightens the carpet against the tack strips.

If the floor is going to be finished with 3/4-in. hardwood strips, I set the side casings on 3/4-in. blocks. When the floor is installed, the blocks come out, and the flooring slips into the gap. □

Jim Britton is a trim carpenter and a contractor living in Fairfield, Calif. Photos by Charles Miller.

15. Undercut the side casing. With the side casing face-side up and slightly tilted, Britton removes material from the miter cut with a bench-mounted disc sander.

16. Now it fits better. By undercutting the side casing with the disc sander, Britton achieves an acceptable miter. A dose of caulk will touch up the remaining crevices.

17. Shim problem casings at the corners. If the casings are out of plane, slip a shim under the corner so that both pieces bear on it. Then trim the shim flush with the casing with a utility knife.

18. Nail 'em. Once the shim has been trimmed, secure the casing corners to the wall with nails driven into the header, trimmer and door jamb. Fill any gaps between the wall and the casing with caulk.

Hanging Interior Doors
An organized approach to a demanding job that must be done right to be done at all

by Tom Law

Among carpenters, door hanging is a prestigious job. It requires skill, patience and a thorough understanding of the steps involved. I spent years hanging doors and fitting hardware when I was employed in commercial work. But for the last ten years I've been a general contractor, and hanging doors is still something I like to do. On those unhappy occasions when speed and economy are more important than quality, I'll install prehung doors, but even then prefer not to. Being able to hang doors properly gives you considerable flexibility in choosing the materials you use for door, jamb, stop and casing. And where custom materials are specified, you may have no other choice than to hang your own.

I generally hang doors in place, even though it costs more in terms of time and money. It gives me something constructive to do on site when I'm coordinating the various trades during the finishing stage. And my clients are assured of properly fitted doors that won't bind during wet weather and won't show large gaps between door and jamb during the winter.

On the average, it takes me about three hours to hang an interior door. This includes setting the jamb, fitting the door, mortising the door and jamb for hinges, hanging the door, applying the stop and casing and installing the lockset. If I have several doors to hang, I don't go through this sequence on each door; rather I move from door to door, repeating the same operation on each, until all are finished. More about this later.

Setting the jambs—Interior jambs are made of clear lumber. They need not be especially strong as the weight of the door is not borne by

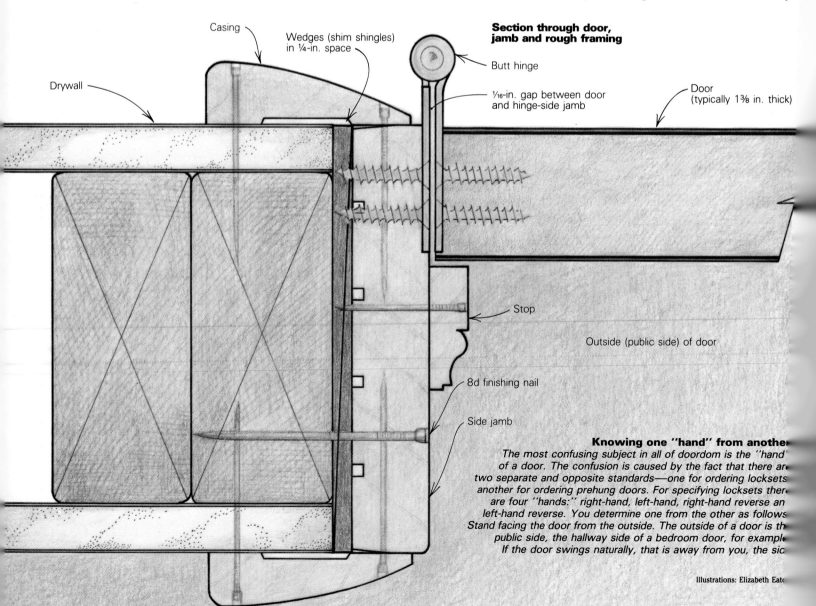

Knowing one "hand" from another
The most confusing subject in all of doordom is the "hand" of a door. The confusion is caused by the fact that there are two separate and opposite standards—one for ordering locksets another for ordering prehung doors. For specifying locksets there are four "hands:" right-hand, left-hand, right-hand reverse and left-hand reverse. You determine one from the other as follows Stand facing the door from the outside. The outside of a door is the public side, the hallway side of a bedroom door, for example If the door swings naturally, that is away from you, the sid

Illustrations: Elizabeth Eato

the jamb but is transferred to the jack studs through the shim wedges (drawing, below). The wood should be straight-grained and free of warp. In the old days, the back side of the jamb was kerfed longitudinally to relieve any stress that might develop and cause the jamb to cup. These days, however, those kerfs have been reduced to shallow V-grooves, which will do little to control cupping and twist. Also, some new jambs are not continuous lengths of lumber end to end, but are made up of finger-joined pieces of clear pine. These jambs are acceptable if painted, but if you want to finish them naturally, you'll want to select clear, solid stock.

The jamb should be slightly wider than the thickness of the finished wall. For example, a nominal 2x4 stud wall with ½-in. drywall on each side should measure 4 9/16 in. thick, and the standard jamb thickness for this wall is 4 5/8 in. Walls are never perfect. They are either too thick or too thin, and it's a problem to make the door jamb and its casing fit perfectly. The head jamb is housed in dadoes cut into the side jambs, which means that an allowance must be made for the combined dado depth when cutting the head to length.

You should take great care when setting jambs to make them as accurate as possible and thereby eliminate unnecessary work when fitting the door. The first thing I do is make a spreader out of scrap ¾-in. plywood for the bottom of the jamb. The spreader's job is to hold the side jambs parallel, so its length should be exactly the distance between the side jambs at the head. It should be about 1 in. wider than the jamb.

After nailing the side jambs and head together and before standing the assembled jamb in place, I measure the rough opening. It ought to be about 2 in. larger all around than the door. This will leave about ¼ in. on each side of the jamb for shimming, assuming that the jambs are ¾ in. thick. A shim space this size is sufficient for plumbing the jamb in a rough opening, but it's a good idea to check the jack studs (also called *trimmers* or *cripples*) for plumb before setting the jamb. If they are out of plumb (more than ¼ in. top to bottom), you'll want to get them right. Next, stand the jamb in the opening, put the spreader in the bottom and temporarily shim between both sides at the top to hold it in place. (For shims, I use undercoursing-grade shingles and rip them on my table saw to about 1½ in. wide.)

Now try the head for level (photo next page, top left). If the finish floor will be carpet, the side jambs can rest directly on the subfloor. All you need to do is shim up under the low side to level the head jamb. But if the jamb has to sit on top of a finish floor like oak or tile, you begin by sawing off the bottoms of the side jambs in an amount that equals the thickness of your finish floor. Then set the jamb in the opening, block up the sides with a pair of scraps that equal the thickness of the finish floor, and shim up one side to get the head perfectly level. Now set your scriber to the distance of the shim space and scribe off the opposite leg, as shown in the photo next page, center left. Remove the jamb, trim off the scribe and replace the jamb with both legs on the blocks. The head is now level.

Place the spreader between the sides at the bottom, and again wedge the top firmly in place. The square ends of the spreader are used to bring the jamb sides into proper alignment. I temporarily wedge the bottom of the side jambs against the spreader. Next, I find the center of the head and mark it on the edge for a plumb line, drive in a 4d finish nail just to the side of the mark and hang a plumb bob on it; this way the center of the string coincides with the center of the pencil line. Then I draw a centerline across the face of the spreader. When hanging the bob, I adjust the line so that the point just misses the wood.

Plumbing and squaring the jamb is now just a matter of bringing the centerline on the spreader directly under the tip of the bob. Do this by loosening the shim shingle on one side of the jamb and tightening the one on the opposite side, as shown in the photo next page, bottom left. Next you should drive in the wedges so they are snug, but before nailing the jamb to the jack studs, lay a short straightedge against the wall and jamb edge to make sure that the jamb projects equally on both sides of the wall.

Now you can nail one side near the top with

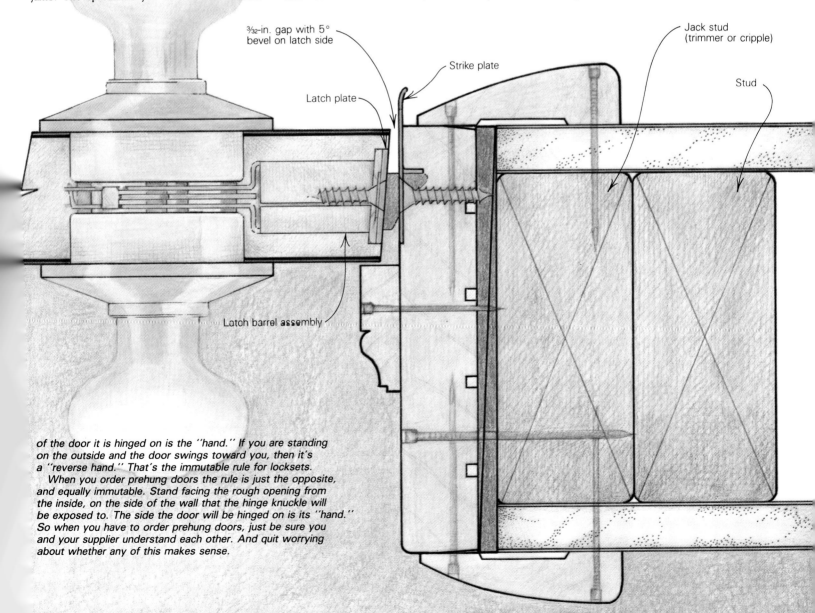

of the door it is hinged on is the "hand." If you are standing on the outside and the door swings toward you, then it's a "reverse hand." That's the immutable rule for locksets.

When you order prehung doors the rule is just the opposite, and equally immutable. Stand facing the rough opening from the inside, on the hinge side of the wall that the hinge knuckle will be exposed to. The side the door will be hinged on is its "hand." So when you have to order prehung doors, just be sure you and your supplier understand each other. And quit worrying about whether any of this makes sense.

Setting the jamb. After standing the assembled jamb in the rough opening, try the head for level, as shown. Then adjust it if necessary by shimming up the side jambs.

Here the side jambs need to be trimmed to rest on oak flooring. A piece of scrap equal to flooring thickness is placed beneath each side jamb as level adjustments are made with a scribe. When the scribe is sawn off and the jamb positioned in the opening, the head will be level.

To get the side jambs parallel, hang a plumb bob from the center of the head jamb and shim the side jambs right or left to get the bob centered over the spreader at the bottom.

Finally, to straighten the sides, use a long straightedge against the jamb and drive in wedges (shim shingles) to bring the jamb into alignment. Be certain to position the wedges at hinge locations on the hinge side.

two 8d finish nails. Nail right through the shim wedges. Check the other side for alignment and nail it. Now go to the bottom, and nail both sides. At this point the chances are that the side jambs won't be straight top to bottom, but will be bowed slightly in or out.

To correct this condition, lay a straightedge (one just shorter than the height of the opening) against one side of the jamb. Lean against the straightedge with your body, and wedge a pair of shim shingles between the jamb and jack stud to bring the jamb out to meet the straightedge (photo facing page, right). Hold the jamb against the wedges and remove the straightedge; give the wedges just a hair more push, and then nail the jamb right through them with two 8d finish nails. The additional push of the wedge moves the jamb slightly out of alignment, but the nailing will compress the small wood-to-wood air spaces, and bring the jamb back into line. Return the straightedge to the jamb and adjust where necessary. Once both jambs have been straightened, they will be parallel.

On the hinge side, be sure to put the wedges near the hinge locations. This will hold the jamb firmly when it is mortised for the hinges and will transfer the weight of the door directly to the wall framing.

After the jamb is nailed up, you can use a handsaw to cut off the protruding wedges, or score them on both sides with a utility knife and snap them off. Be sure they are behind the wall surface so they won't be in the way of the casing when it's applied.

The next step is applying the door stop. Most interior doors are 1⅜ in. thick, so use a combination square and a pencil as a marking gauge to scribe a line 1⅜ in. in from the latch side of the jamb. On the hinge side, this line should be 1⁷⁄₁₆ in. in from the edge of the jamb. The additional ¹⁄₁₆ in. is necessary to prevent the door from binding or stopping at the end of its swing.

The corners of the door stop can either be mitered or coped; coping is usually better. As a rule, I cut the stop and tack it in place, but I don't nail it up until after the door is hung because it may be necessary to make adjustments to get the door to close properly. Also, I have to remove the stop on the hinge side to accept the hinge-mortising templates. The casing can be applied now or after the door is hung.

Fitting the door—Gone are the days when a heavy door jack was used to hold the door on edge while a carpenter shaved it to size with a jointer plane. While it's fun to swish off 7-ft. long curls of pungent pine with a hand plane, using a power plane is faster, and it takes much less effort. Instead of a door jack, I use a door bench with enough room on it to hold the tools I use— a power plane, a router for hinge mortising, a skillsaw and cutoff guide, a belt sander, a drill, a jack plane, a Stanley No. 271 router plane, which I use to mortise for latch plates and strikes, and my Yankee screwdriver.

I've known carpenters who could place a door in an opening, mark all the margins and mark the hinge and lock locations in one trip; then they could cut the door to size, mortise for hinges and lockset and return it for hanging without further adjustments. I'm seldom that lucky, and the method I use allows for a trial fitting and a final trimming.

The traditional margins for a door are ¹⁄₁₆ in. of space between jamb and door at the head, ¹⁄₁₆ in. on the hinge side and ³⁄₃₂ in. on the latch side. Old carpenters used to call it nickel and dime—a nickel on each side and a dime on top. They used the coins as feeler gauges; a nickel is a big ¹⁄₁₆ in. and a dime is a thin ¹⁄₁₆ in. Since ¹⁄₁₆ in. is the smallest graduation on a carpenter's rule, thirty-seconds are referred to as big and little sixteenths in carpenter-speak. The old rules make a door a little too tight in my opinion, so when I fit doors, I use a nickel on top, ¹⁄₁₆ in. on the hinge side and ³⁄₃₂ in. on the latch side. The space at the bottom is from ⅛ in. to ½ in., or large enough to clear any high spots on an out-of-level floor.

Doors, like other things made of wood, shrink and swell with changes in season. The part of the country where I live has wide swings in humidity and heat from summer to winter, and the time of year has to be taken into account. A door fitted tightly in winter will strike the jamb when closed in summer, and a perfectly fitted summer door will be drafty in winter. In the summer, fit close; in the winter fit loose. It's a common mistake to recut the margin on a door during its first summer, only to have it too small when winter returns. If a door must be planed down, reduce it only enough to clear. Also, don't plane the latch side—if you do, the lockset will need to be shifted and some of them can't be moved without exposing the holes. Plane the hinge side, then deepen the mortises and reset the hinges.

Fitting a door must proceed step by step. Lay the door across the top of the bench or on a pair of sawhorses, and cut it just small enough to fit snugly in the opening. If you have to remove more than a ⅜-in. wide strip, use your skillsaw with a sharp planer blade. If you have to saw off the bottom of a veneered door, first score the line of cut with a sharp utility knife. This will keep the veneer from tearing and chipping out. But if you're taking the edges of the door down by ¼ in. or less, it's easier to use a power plane (the direct-drive type is designed for edge planing). After trimming the door to fit the opening, place it in the jamb and force it up against the head by wedging from the floor and over against the hinge side by wedging from the latch side. Scribe a line across the head ¹⁄₁₆ in. below the jamb and another line the required distance up from the floor.

On the latch side, measure over from the jamb ³⁄₃₂ in. or a fat ¹⁄₁₆ in., and mark it on the door. And on the hinge side make a mark ¹⁄₁₆ in. over from the jamb. Remove the door from the jamb, set your combination square by the mark

The edge on the latch side of a door must be beveled about 5° if it is to close without striking the jamb. This can be done with a hand plane or by using a power plane, as shown at right. The fence is set about 5° off the perpendicular, and the depth of cut is set for about ¹⁄₃₂ in. Successive passes are made until the scribed line on the open side is reached.

on the latch side and mark a line down the entire length of the door. Flip it over and do the same on the hinge side. Secure the door on edge and plane to the line on the hinge side and on top and bottom.

Now you can trim and bevel the latch side of the door. Flip the door over so the latch side is up and the scribed line toward you. This edge must be planed to the line and beveled. As a door swings on its hinges, the leading edge will strike the jamb if it's not beveled. The degree of bevel can vary, but I find that 5° is good in most cases. So when setting the fence on my power plane, I adjust it by eye about 5° out of square. I set the plane's depth of cut to about ¹⁄₃₂ in. and make successive passes from one end to the other until I shave off the top of the line.

Return the door to the jamb, wedge it in place and check the margins on all four sides. If any side needs a slight trim, plane it down and check again. Once the fit is correct, I take the arris off all of the edges of the door with my smooth plane. Sharp 90° edges on a door can be dangerous.

Mortising for hinges— To cut hinge mortises in the door and the jamb, I use a router and a pair of site-made templates connected by a bar. This setup eliminates the need to lay out the hinge locations on each door and every jamb, and it makes for quick, accurate work. This can be especially important if you have a houseful of doors to hang. But if you're hanging only one or two doors, you might want to cut the mortises with the chisel in the traditional way that's described in the sidebar on p. 59. Manufactured hinge-mortising templates are available, but I prefer to make my own because it's less expensive, offers me greater flexibility and reduces the amount of permanent gear and dinky parts I have to keep track of and take from job to job. All that's required using my method is a sharp butt chisel, a router, a ½-in. straight-face bit (preferably carbide tipped) and a ⅝-in. O.D. guide bushing, sometimes called a template guide, for the router.

I begin by laying out a single door and its jamb for hinges. Place a nickel on top of the door and wedge it in place (you can do this at the same time you check the margins for accu-

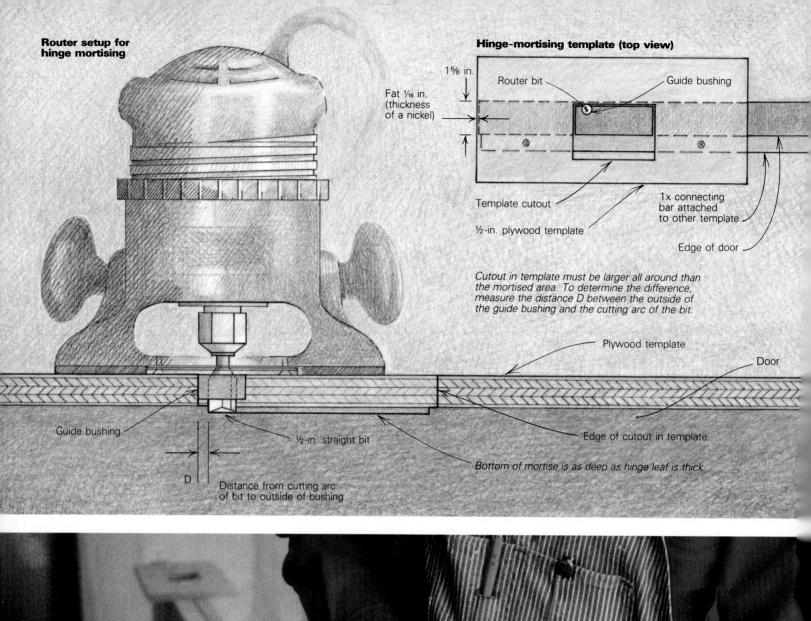

racy); then use a straightedge to strike a line across door and jamb simultaneously. Then mark the side where the hinge will be. That seems simple enough, but many a hinge has been confidently mortised into the wrong side of a door. Mark the location for the top of the top hinge and the bottom of the bottom hinge, and the center of the center hinge if your door has one. The top of the top hinge should be 5 in. from the top of the door, and the bottom of the bottom hinge 5 in. from the bottom of the door.

With the door on edge, hinge side up, draw the lines you marked on the face across the full edge of the door. Position one of your hinges on the correct side of the line so that the center of the barrel or knuckle is about ½ in. from the edge of the door. This approximation is for standard 4x4 butts; distance will vary slightly depending on door thickness, hinge size and clearance needed to keep the door from contacting the casing when opened a full 180°. Positioned properly, the back edge of the hinge will lie about ³⁄₁₆ in. from the back edge (the closed side) of the door. Holding the hinge firmly on the layout line, trace around the back edge and opposite side with a sharp pencil, or score lightly with your penknife.

Next, cut a pair of rectangles about 14 in. long and 6 in. wide from a scrap of ½-in. plywood. These will be your router templates, and into them you'll jigsaw rectangular cutouts that will guide the bushing on the router to produce a precise mortise in the edge of the door (and jamb) that will hold the hinge leaf.

To get the mortise the right size, the template opening will have to be larger than the hinge outline by a specified amount. As shown in the drawing, facing page, this amount equals the distance from the outside of the guide bushing to the cutting arc of the bit. So to lay out the template, trace around the hinge leaf, and enlarge this outline on all three sides in the correct amount. Lay out the templates for both hinges so that the top edge of the finish cut for each is 5 in. from the outer edge. Leave enough room on the open side of the template to maneuver the bit up to the edge of the door. Now saw just shy of the line, and then pare right down to it with a sharp chisel.

Mount one template over a trial hinge layout on the edge of a length of 2x. Secure the template with a couple of 4d finishing nails; then make a trial cut with the router.

Now remove the template (leaving the nails in their holes) and square up the radiused corners of the test mortise with your chisel. Place the hinge in the mortise to check for a snug fit. If it's too tight and the hinge must be forced in with a hammer, enlarge the cutout in the template slightly with a bastard file and check the fit again. If the hinge is too loose it may not stay square, which will cause it to be out of alignment with its mate on the jamb. To remedy this condition, you can add veneers of masking tape to close up the space.

The depth of the mortise should be exactly the same as the thickness of the hinge leaf. The old-timers used to say, "It looks like it growed there." But there are some instances, which I'll discuss further on, when the hinge depth needs to be varied. To set your router bit the proper depth, turn the router upside down, and gauge the depth of cut by holding the hinge leaf on top of a small scrap of ½-in. plywood. Make another trial cut to make sure that the hinge will fit flush with the wood around it.

With the two templates complete and the router bit set to the proper depth, position the top-hinge template over the hinge layout on the edge of the door. Because the template for the top hinge will fit right against the head when you mortise the jamb, it needs to be flipped end for end and then offset when you mortise the door to produce the margin between door and head. As shown in the photo, facing page, I hold a nickel on the top edge of the door and bring the top edge of the template out flush with it to get the correct spacing. But make sure that the back of the cutout is positioned parallel to the layout line at the rear of the mortise, and that the spacing between templates is correct. Now nail the second template over its layout at the bottom of the door.

Now get a straight 1x ripping that's about as long as the door, and attach it to the underside of both templates (with drywall screws and glue), making certain that it fits closely against the edge of the door. This bar serves to hold the two templates the proper distance apart throughout the whole job, and it acts as a fence, ensuring that each mortise is the correct width by holding the back edge of the cutout at a constant distance from the edge of the door.

After mortising the door, take the template to the hinge side of the jamb, index the top of the top-hinge template hard against the head jamb and tack it in place, top and bottom. Then rout the mortises (photo below right).

Installing the hinges—When the routing is done and you've squared up the inside corners with your chisel, place the hinges in the door mortises (make sure barrels are oriented pin up) and mark the centers for the screw pilot holes. For this I use a nail set because it's blunt and crushes its way into the wood fibers. A sharp punch or an awl can be pulled off center by coarse grain. I don't make the pilot hole exactly in the center of the screw hole, but just a hair off center, in toward the back (closed end) of the mortise. Doing this will pull the hinge leaf snugly against the back wall of the mortise when the screws are driven home.

Phillips-head screws are better for hinges than slotted screws. The Phillips screws can take a lot of torque without letting the bit slip out of the slot and burr the head of the screw and scar the hinge leaf. Door men use a Yankee screwdriver because it saves lots of time, keeps your hand from aching and your forearm from cramping. To ease the friction while driving, drag the screw threads though a bar of wax made by melting and mixing paraffin and beeswax. The spiral on a Yankee screwdriver requires lubricating from time to time, but don't use oil, which will splash on the work, get on your hands and wind up everywhere. Paraffin is good for this and it won't make a mess.

Hanging the door—Now that the door is trimmed to size and the hinges mounted, it's time to hang it on the jamb. One method is to split the hinges and with one leaf on the door and its mate on the jamb, lift the door, slide the knuckles together and insert the pins. This works fine if the door is light, if there are only two hinges and if their alignment is perfect. If there are three hinges and their alignment is just less than perfect, hang the door on the top and bottom hinges and then mount the center hinge in place with the door open.

Close the door and check the margins—they should be perfect all around, ³⁄₃₂ in. on the latch side, ¹⁄₁₆ in. on the top and ¹⁄₁₆ in. on the hinge side. There are, unfortunately, occasions when this observation causes your heart to sink. All that careful work and look at it. The top's not parallel to the head, and on the latch side it's closer on the top than at the bottom, and on the hinge side it's closer on the bottom than the top. Despite your despair, this condition is not unusual, and it can be corrected.

Heavy doors can sag once they're hung because they take up the slight play between the knuckles and pins. To compensate for the sag, the hinges must be adjusted. There are several ways to do this. You can pull the top in closer by making the mortise in the door just a little deeper than the leaf thickness. Or you can push the bottom out by shimming behind the bottom hinge with a piece of compressed cardboard. (The box the hinges came in is about right.) Remove the bottom leaf from the jamb and insert a leaf-size piece of cardboard behind it. Repositioning the hinge over the cardboard will push the bottom of the door over and move the top edge up. Usually this is enough to take the sag out, but if it isn't, you can pull the top over by

Mortising the door. Law tacks site-made hinge-mortising templates to the edge of a door (facing page). He uses a nickel as a feeler gauge to position the template a fat ¹⁄₁₆ in. beyond the top of the door.

Mortising the jamb. Because the template will fit hard against the head when the jambs are mortised (right), the offset on the door will produce the proper gap between door and head jamb. The whole jig has to be flipped end for end when routing the jamb, so the layouts have to be symmetrical. Before routing the jamb, make sure shim shingles are directly behind the mortise to transfer the weight of the door to the studs.

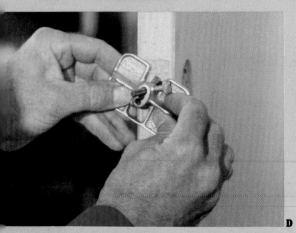

Installing a lockset

Most interior locksets consist of three parts—a pair of knobs, a latch assembly with integral latch plate and a strike plate, which is mounted on the jamb. To install these three items you bore three separate holes. They must all relate to one another in a precise and prescribed way. Some locksets come with templates that help you locate the three bore centers; others come only with instructions that tell you where to position the holes relative to the edges of door and jamb.

You begin by laying out the bore centers for the knobs and the latch-barrel assembly. Then you lay out the center for the strike-plate bore on the jamb. To locate the center for the knobs, measure up from the bottom of the door 36 in. to 39 in. (depending on your preference) and mark a light line with your combination square across the edge and both faces of the door. Then, using the square as a depth gauge, mark a vertical line 2⅜ in. or 2¾ in. (depending on the lockset you have) back from the edge of the door. Next find the center for the latch-barrel bore by dividing the line on the door's edge precisely in two. Finally, mark the center for the strike-plate bore by finding the point on the jamb exactly opposite the center on the door for the latch bore.

After punching all these centers, the next step is to cut the hole for the knob assembly (A). I do this with a 2⅛-in. hole saw (standard for most locksets), though you can use an expansion bit in a hand brace. Saw about halfway through on one side and then finish the hole by cutting from the other side. Next, chuck up a ⅞-in. dia. spade bit in your drill and bore the hole for the latch barrel (B). Be sure to hold the bit level so that the bore breaks through into the center of the knob hole. Next, using the same bit, bore out a ⅝-in. deep hole in the jamb for the strike plate.

Now you can lay out and cut the shallow mortises for the latch plate and the strike plate. For this I use a 1-in. chisel and a little Stanley No. 271 router plane. Begin by inserting the latch barrel into its hole and squaring up the plate with the edge of the door. Trace around it carefully with a sharp pencil or a striking knife. Remove the latch, and cut the wood at the top and bottom of the layout by giving the chisel a smart tap with your hammer (C).

Then do the same along the sides, though you have to be careful here not to drive the chisel too deeply and risk splitting out the wood along the edge of the door. Set the cutter on the router plane to the thickness of the latch plate and remove the wood inside the mortise layout (D). With a router plane you can quickly excavate the mortise to a uniform depth all around. The router plane also ensures that the bottom is flat. Repeat this procedure on the jamb for the strike plate, and you're ready to install the latch (E). —*T. L.*

mortising a little deeper or by "throwing" the hinge to move its pivot point.

Looking down at a section drawing taken through the door, hinge and jamb, the center line of the margin passes through the center of the hinge pin. "Throwing" this pin center left or right has the effect of moving the door left or right. To do this use cardboard shims as before, but this time cut them into narrow strips about ¼ in. wide. By placing one of these strips between the jamb and leaf at the rear or closed side of the mortise, you will throw the hinge to the right. Putting a shim strip on the open side of the mortise will throw the hinge to the left.

Quicker by the dozen—When I'm hanging a house full of doors I follow this routine: First I sort the doors and jambs for size or type. Then I cut the jamb heads to size, nail them to the sides, and take each assembed jamb to its rough opening where I lean it against the wall. Generally I make the jamb width the nominal door size, such as 24 in., 30 in., or 36 in. But you should be sure of the door size, because some of them are already cut down. Using the technique described above for setting the jambs, I start at one end and go to one jamb after another until they're all set. Next, I cut the stops and tack them in place. Then I start at the beginning again, and using the plywood router template, I mortise all the jambs for hinges.

At this point I have not applied any of the casings because they get in the way of the mortising template. Also, if any adjustments to the jamb are needed after the door is hung, I can still get to the wedges.

With the door on the bench and latch side up, I plane it to size and cut the bevel. Since I know the size of the opening and know the sides are straight, I just plane the door to 5/32 in. less than the width between the jambs on the bench. Then I flip it over and rout out the mortises. Then I install the hinges, take the jamb leaf to the jamb and mount it, carry the door to the jamb, hang it on the top hinge, open the door and drive the screws on the bottom hinge. I close the door and check the margin. No heartaches. The casing can now be applied. The next step is to go through and drill out and mortise for the locksets, and then installing them. When the doors are hung and locksets installed (sidebar at left), the stops can be adjusted to fit those doors that are slightly warped.

By moving from door to door with each operation, you can save a lot of time because you have each tool you need for each repeated operation. Using this production-line method, I can approach the speed of installing prehung doors, but still my way costs more. Is it worth it? I think that it is, because I can achieve better results than what comes from the mill. Many of the assembly-line workers in mills do shoddy work, and quality control is down. I've seen every possible mistake in doors and jambs that come as units from the mills. Often it's easier to do the job right yourself than to fix what another person has fouled up. □

Consulting editor Tom Law is a carpenter in Davidsonville, Md.

Casing a Door
Work carefully, and save the wood filler for the nail holes

by Bob Syvanen

Nowhere in house building is the workmanship more obvious than in interior trim, and doors are among its most visible locations. It takes skill, patience and the proper tools to do good work. Perhaps more important, the use of some special techniques will lead to results that the craftsman can be proud of.

The wood most commonly used for trim is pine, though almost any wood will do—solid wood, that is. Various alternatives to solid wood are now available. Such materials as plastic and hardboard are more stable than solid wood and less likely to split. But while they may look like wood, they don't feel like wood, smell like wood or work like wood. There is really no substitute for the real thing. If you're going to work with the best, use #1 clear material.

Casing stock often comes in random lengths from 7 ft. to 20 ft. Usually, however, you'll find it in the standard lengths of 10 ft., 12 ft., 14 ft. and 16 ft. You can sometimes purchase complete trim packages for standard-sized windows and doors. A door package will include two head casings (the horizontal pieces above the door) and four jamb casings or side casings (the vertical pieces at the sides of the door).

Door casings come in various widths and profiles, but all are partially relieved, or plowed, on the back surface. (For a useful booklet on wood molding and casing patterns, contact the Western Wood Moulding and Millwork Producers Association, Box 25278, Portland, Ore. 97225.) The relieved section is about 1/16 in. deep, and leaves a shoulder about 3/8 in. wide on each side of this trough. When the trim is installed around the opening, one shoulder rests on the door frame and the other rests on the wall. The relieved section bridges any high spots that might be between them, and allows the casing to make good contact with each.

A quick way to relieve shop-made casings is to run the stock on edge through a table saw, with the blade set at a slight angle. If you turn the stock end for end and repeat the cut, a triangular section from the back side will be removed, leaving two flat shoulders like those on mill-cut casings. The drawing at right shows three styles of casing you can make in the shop.

If the head casing is to be square cut and laid on top of the side casings, it can't be relieved along its entire length because the relieved area will show up at the ends of the head casing. On such pieces, relieve the trim using a dado blade on your table saw. Start and stop the blade just short of the casing ends.

Tools—Any tool that makes a job easier will usually improve your work, since it will let you concentrate more on the result and less on working the tool. The best example of this I can think of is my miter box. Years ago, I used a homemade wood miter box. When I switched to a heavy-duty metal miter box with its own backsaw, I did better work faster. Now I use a power miter box. It speeds up the work and allows me even more time to concentrate on what I'm doing. (Power miter saws are sometimes called chop saws.) Now I wouldn't think of trimming out a house without one. The quality of a tool also makes a big difference. Good work cannot be done with a dull saw, a dull plane or a power tool that makes sloppy cuts.

In addition to the power miter box, the tools I use for casing a door are these: an orbital sander, a 13-oz. hammer, a block plane, a 2-in. and a 1/2-in. chisel, a small and medium nail set, a combination square, a framing square, a utility knife and a tape measure.

I also consider my portable work stool to be standard equipment for interior trim work. This stool, called a cricket, is simply a step made out of 3/4-in. material, with a shelf below to hold tools, nails, glue and sandpaper. Mine is about 14 in. wide by 25 in. long by 19 in. high, but the measurements can vary depending upon your needs. What's important is that you can stand on it and reach the head casing, and that you have convenient access to your tools. There should be a hand hole in the top so that the whole "casing shop" can easily be picked up and carried to the next opening.

Consider the opening—It's easier to case a door if the rough framing of the opening is square and plumb, with plenty of support for the trim. The thickness of the walls should be consistent, but often it isn't. The jambs may be poorly aligned or of different widths, and the wall finish may be sloppy—a real pain when you're trying to get casing to fit neatly to both jamb and wall. And all too often, sheetrockers will stop their sheets at either side of the rough opening and fill in the space above with a small piece. This usually causes problems. Most important for the finish carpenter is the thick build-up of joint compound at the corners of the door opening. Sometimes the excess material can be sanded down or scraped away, but it's usually difficult to case over.

A successful casing job also depends on how well the door jamb was installed. For casings to fit properly, jambs should be 1/8 in. wider than the finished wall thickness. This leaves the jamb 1/16 in. proud of the drywall on each side of the door, and allows you to bevel the edge of the jamb nearest the drywall (before the jamb is installed) so that the casing will fit snugly.

Casing the opening—The first step in casing the opening is marking the jambs for the setback, or reveal. The reveal is measured from the inside face of the door jamb (drawing, next page), and allows for a slight margin of error when installing the casing. I make the reveal about 1/8 in. to 1/4 in., depending on casing size and type, but it's mainly an aesthetic decision.

A combination square, set for the depth of the reveal and used as a gauge, can be used to mark the reveal. It works quite well, but I prefer to make up a wood gauge block. A gauge block won't lose its adjustment, as a combination square sometimes does. I make my block from a 2-in. by 4-in. piece of 3/4-in. thick scrap stock, with a rabbet cut in it on four sides to the depth of the reveal. To mark the reveal, I lap an edge of the block over the door frame, and run it along the frame to guide my pencil. The resulting line is quite accurate. The gauge block also makes it easy to mark the reveal at the corners

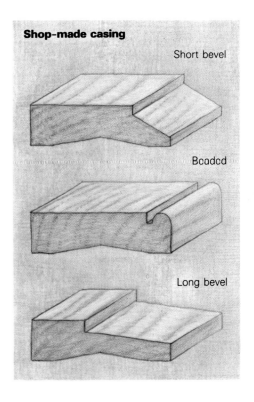

Shop-made casing

Short bevel

Beaded

Long bevel

of the door jamb where a combination square can't quite reach. When two people are installing casings, it's a good idea to cut identical gauge blocks for each so that all reveals will be the same.

There are three basic styles of door casings: square-cut, mitered and corner block.

Square-cut casings. Simple square-cut casings (drawing A, facing page) are fairly easy to cut and install. Once the reveal has been marked, the casings can be cut to rough length and the pieces distributed to appropriate openings. I like to do this in order to be sure I have them all. It's upsetting to interrupt the flow of work in order to cut, shape or order one extra side casing. Working one opening at a time, I begin fitting the casings by making a square cut at the bottom of each side casing, checking each one for a precise fit at the floor after aligning it with the reveal line. When the casings must be installed before the finish floor is laid, I fit them to a scrap wood block as thick as the finish floor will be.

With one of the side casings held in place, I align it with the reveal line on the jamb, and mark it where it intersects the head reveal line. I use a sharp knife to make a precise mark on the inside edge of the casing, not on the face. A square trial cut can then be made about ⅛ in. above this mark, followed by a trial fit with the head casing. What's important here is to determine the correct angle. If the cut isn't quite right, an adjustment can be made when the piece is recut. The second cut is made a tad below the first one, but still shy of the true cut mark. When the angle of the trial cut is perfect, it can be duplicated at the true cut mark. The casing can then be aligned with the reveal line and tacked in place. Drive one 6d finish nail partway into the stud and one 8d finish nail partway into the jamb so that the casing can be removed for more trimming if necessary. Repeat the procedure for the remaining side casings.

The head casing is completed simply by squaring one end, holding it in place on top of the side casings, marking it to length with the knife, and cutting. Some people like to cut the head casing slightly long, so that it overhangs the jambs by ¼ in. or so.

After double-checking the alignment of the side casings, I nail them in place at the door jamb with 6d finish nails (4d for thin trim and some molded trim). Hardwood should be predrilled before you nail it. The casing edge that is against the wall should be nailed with 8d finish nails into the trimmer studs. I also like to toenail a 4d finish nail through the end of the head casing and into the top of the side casing. This helps to keep the joint surface aligned. The trim should fit tightly against the door jamb and the wall, so use as many nails as it takes to do the job. A good way to test the trim for tightness is to rap the casings with your knuckles. If they make a rattling "clack-clack" sound, put another nail in and rap again. When you hear a solid-sounding thump, the casing is tight. All nails should be set.

As an additional precaution against joint separation, I usually run a bead of white or yellow glue along wood-to-wood contact points just before nailing. These are very visible joints, and when glued and nailed, they'll hold together even if the wood shrinks. Wipe all traces of glue off the surface, using a damp cloth. This is very important if the casing is to be stained.

Variations of square-cut casing. One variation uses head casing that is thicker than the jamb casings. It looks best when these head casings extend past the jamb casings by the difference in their thickness (drawing B). For example, if the jamb casings are ¾ in. thick and the head casing is 1 in. thick, the head casing would extend past the jambs by about ¼ in. on each side. A second variation starts with a standard square-cut casing, and then adds mitered backband molding around the outside edge (drawing C). A third variation is used when the inside edge of a flat casing is beaded. This technique requires that you miter the beaded edge, square cut the ends of the head casing and cover the outside edge of the whole casing with a mitered backband molding (drawing D).

Mitered casings. The second, and most common, way to trim out a door is to miter the corners of the casing (drawing E). With mitered casings, a trial 45° cut is made instead of a square cut at the top of the side casings, but otherwise the trial cutting and fitting are the same as with square-cut casing. If the door jamb isn't square (use a framing square to check the corners where the head jamb meets the side jamb) and can't be adjusted, trial cuts on scrap stock will help you find the proper angle for the head casing. The side casing and head casing must have the same angle cuts in order for the outside corners to meet and for the molding profiles to match. This step is very important if the casing will be stained, since the joint is so visible.

When the outside corner of a trial miter joint is open, the angle of the next trial cut will have to be adjusted to take more stock off the heel, or inside corner, of the miter. Remember that head and side casings must be cut using the same angle. The adjusted cut on the head casing, therefore, should reduce the gap by one half, with the remaining half cut from the jamb casing. These adjustments are usually so small that changing the saw angle with any precision is difficult.

To make fine adjustment cuts using a miter box, I put a wedge between the miter-box fence and the stock—pieces of cardboard, wood chips or even plane shavings will work. By moving the wedge away from or closer to the saw blade, slight variations in the angle of the cut can be made. Another way to make fine adjustments is to use a block plane. Make sure the iron is sharp and set for a fine cut.

Mitered casings are installed in much the same way as square-cut casings. The only additional nailing that's required is at the outer portions of the mitered corners of the head and side casing, where 4d finish nails or brads should be sunk from each direction.

Corner-block casings. The third method of trimming out a door is to fit the ends of head and jamb casings to corner blocks (drawing F). The corner block should be a little thicker than the side casing, and can be combined with a similarly shaped plinth block. The plinth block's purpose, aside from making the visual transition between the casing and the baseboard, is to provide a wide, flat surface for the baseboard to die into.

The plinth block is installed first, and should be plumb. It can be installed with a reveal that matches the casing reveal—⅛ in. to ¼ in. is fine. Installing the casing from this point is the same as described earlier, except that the side casings are fitted and butted to the plinth block.

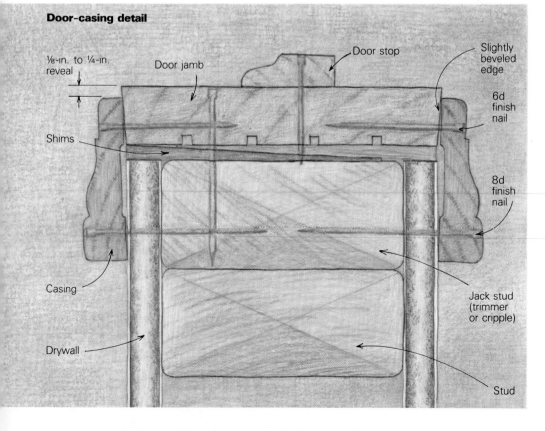

Door-casing detail

Mating the surfaces—These techniques all work well when the walls are flat and evenly dimensioned around the door opening. But since a house isn't usually built to the tolerances of furniture, wall surfaces won't be perfectly smooth, and the openings won't be textbook square and plumb. Fortunately, there are ways to compensate for irregularities.

With square-cut casings, a little tipping away from the plane of the wall can be tolerated, but not much. When the top of a head casing tips toward the wall, the entire length of the joint in front will be open. To correct this, put a wedge-shaped shim behind the head casing to bring it into proper alignment. A spot of glue will hold the wedge in place.

You can also shim mitered casings, but when the tipping is minor, the back side of the 45° angle can be shaved with the block plane to bring the trim into alignment. This sort of adjusting cut can also be made in a miter box by placing a shim between the casing and the bed of the miter box, near the blade. The resulting back cut will allow the front joint of the miter to close up. If the front edge of the miter must be trimmed down, a wedge placed under the casing at a point farther from the blade will do the trick. All these cuts are made long for a trial fit before cutting to exact length.

When the thickness of the wall is greater than the width of the jamb, the casing will tip in toward the opening, resulting in a visible gap between the casing and the wall. The surfaces of the casing will be out of alignment as well. The gap can be eliminated by planing the shoulder of the casing to fit the wall surface. The alignment problem can be a bit more difficult to deal with, but not impossible. Unfinished casings with flat surfaces can be surface planed with a sharp block plane and sanded with an orbital sander or a sanding block. If the trim is to be stained, however, great care must be taken to avoid sanding across the grain, as the marks will show in the finish. Molded casings can be trimmed to align at corners using sharp chisels.

Wherever shims are used or a casing is planed, it's important to adjust the other trim pieces that will be affected. For example, if a side casing is thinned at the top, the head casing must be thinned to match. If the side casing is thinned at the bottom, then the adjoining baseboard must be thinned, as well.

Filling nail holes—I find that painted casings require more effort to prepare for than stained casings. Paint accentuates any imperfections, and slight dimples left by partially filled nail holes show like beacons through the paint. So the filling must be done carefully. For tools, you'll need a putty knife, good eyes and a sensitive hand.

The best filler I've found was recommended to me by a painter. ONETIME Spackling (Red Devil, Inc., 2400 Vauxhall Rd., Union, N. J. 07083) is an acrylic spackle, and its smooth, putty-like consistency makes for easy application. It dries quickly and sands easily, so you can paint over it almost right away. Using the putty knife or a dab on your finger, fill the holes as completely as you can (if you've done the casing correctly, you won't have any joints to fill). After the spackle dries (a couple of minutes), use fine sandpaper to smooth away the excess. Check the result by rubbing the filled areas with your hand—you'll feel imperfections you can't see. Spackle again where necessary, sand and check. If the first coat of paint reveals a nail dimple, reach for the spackle again.

When the job calls for the casing to be stained, filling nail holes goes a bit differently. I find it's easier to stain the casing before I install it since it's easier to do the job at my workbench. After the casing is installed and the nails are set, I brush on a coat of polyurethane. This protects the wood, and will keep the putty I use from staining the wood around the nail hole. Once the polyurethane has dried, I fill the nail holes with Color Putty (Color Putty Co., 1008 30th St., Monroe, Wis. 53566). It comes in many colors to match various stains, and you can also mix colors to match unusual stains. I rub it into the hole with a putty knife or the back of my thumbnail, and remove the excess with a rag or my finger. After the putty has dried, I apply at least one more coat of polyurethane to the casing.

Refinements—Before finishing, all corners and edges of the casing should be eased. Sharp edges don't hold paint well, and they dent easily and are uncomfortable to touch. I like to block-plane these edges before installing the casing, using one continuous stroke for each cut. A continuous stroke leaves a smooth surface that keeps the same bevel along its length. If you have to plane an installed casing, you'll need a bullnose plane, a chisel or sandpaper to get at the inside corners.

Occasionally a piece of casing won't look as good as you think it should, so you have to remove it. If the piece has to be used again, the nails should be pulled through from the back-side of the wood. This is particularly important when working with prestained trim. If you drive the nail out head first, it will tear the surface of the wood as it exits. To pull nails out from the backside, use either a hammer with sharp edges on the claw or pincer-type pliers. The hole can easily be filled later.

When I'm through casing a door, I load my tools on the cricket shelf and move the whole thing to the next opening. I never leave an opening until it's completely finished and ready for the painter—it's too easy to forget something otherwise. □

Bob Syvanen, of Brewster, Mass., is a consulting editor of Fine Homebuilding *magazine.*

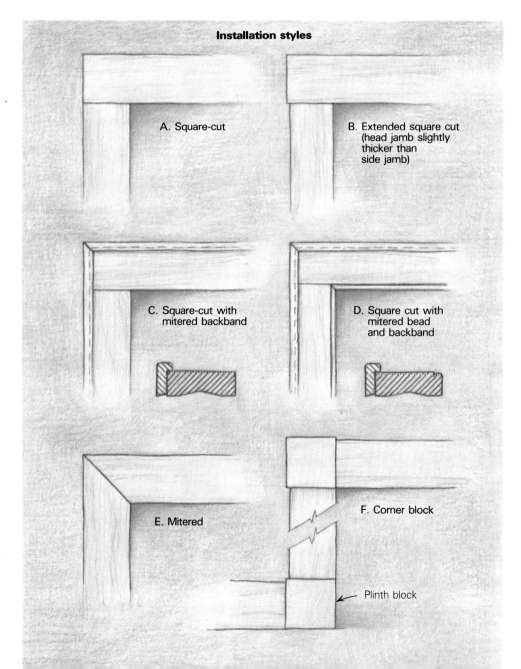

Installation styles

A. Square-cut

B. Extended square cut (head jamb slightly thicker than side jamb)

C. Square-cut with mitered backband

D. Square cut with mitered bead and backband

E. Mitered

F. Corner block

Plinth block

Building Interior Doors

Using a shaper to produce coped-and-sticked frames and raised panels

by Joseph Beals

The dark, tired hallway in our Cape Cod-style house badly needed a face-lift. Like the rest of the interior, the hall had been built on a tight budget. With its aging acoustical-tile ceiling, cheap plywood paneling, assortment of ill-fitted stock trim moldings and five pre-hung hollow-core doors, there was little worth saving. My helper and I gutted the hall of everything but the existing door jambs and vinyl flooring. To give an illusion of height while creating an eye-catching detail, we installed a vaulted ceiling using drywall and curved 2x6 scabs fastened to the ceiling joists. We also applied ½-in. drywall to the walls.

But the heart of the hallway remodel was the construction and trimming out of five solid-wood, frame-and-panel doors (photo right): two 30-in. wide bedroom doors, a 30-in. wide bathroom door, a 34-in. wide cellar door and an 18-in. wide closet door. Four of the five existing doors had been hung on split jambs, with the adjustable portion of the jambs located on the hall side. This enabled us to reset the jambs flush with the new drywall. The closet door was hung on a conventional jamb, which we removed and reinstalled flush with the drywall.

Picking the pattern—The typical commercial frame-and-panel door is the six-panel Federal-style whose almost-flat panels are raised in the most minimal sense of the word. This generic reproduction style would be out of place in our remodel. Rather than subcontract the manufacture of custom doors to a local shop, I decided to design and build my own.

I drew principally on two references for design information: a turn-of-the-century English work called the *Handbook of Doormaking, Windowmaking and Staircasing* (reprinted by the Sterling Publishing Co., Inc.; 212-532-7160); and "Making Period Doors," an article that appeared in *Fine Woodworking* magazine (*FWW* #71, June/July 1988, The Taunton Press, Inc.). After sketching several options, I designed a four-panel Greek-Revival door with panels raised flush to the frames on both sides. This not only creates a very strong play of light and shadow, but it allows the convenience of a single thickness of stock for all door parts. For the master-bedroom door at the end of the hall, I substituted two round-top, leaded-glass panels for the top two raised panels. The narrow closet door has just two raised panels, one above the other.

Although there would be three different door sizes, the dimensions of stiles, rails and mullions would be consistent throughout (drawing facing page). I made the mullions ½ in. narrower than the stiles, and the middle (or *lock*) rails ½ in. narrower than the bottom rails, to give a balanced appearance. The centers of the lock rails are located 36 in. from the bottoms of the doors to produce classic Greek-Revival proportions.

The construction of the doors would echo that of the typical commercial frame-and-

A classical face-lift. Trimmed with custom casings, this shop-built, frame-and-panel door is one of five new doors that highlight the author's hallway remodel.

panel interior wood door—that is, cope-and-stick joinery reinforced with dowels where rails meet stiles. Coping and sticking involves the cutting of a continuous decorative molding (the *sticking*) along the inside edges of the stiles, rails and mullions, then coping the ends of the rails and mullions so that the end cuts are a perfect inverse of the sticking and fit snugly against it. This method surrounds each door panel with molded edges that appear to be mitered at the corners.

A century ago, cope-and-stick joints were cut using a variety of molding planes, chisels and gouges. Now, they're typically produced with a shaper. I own a ¾-in. shaper and a set of carbide cope-and-stick cutters with an ovolo molding profile on them—just what I needed for this job.

The choice is poplar—Most commercial paint-grade interior doors are made of pine or fir. I built mine out of 8/4 poplar planed to a thickness of 1⅜ in., a standard dimension for interior doors. Poplar is a relatively stable, straight-grained hardwood that's harder than most softwoods, works easily and takes a painted finish extremely well. Poplar isn't particularly rot-resistant, however; I wouldn't recommend it for exterior doors.

The poplar cost $1.40 per bd. ft. I don't own a thickness planer, so my local supplier planed the poplar for me for an extra 15¢ per bd. ft. To reduce waste and help cut costs, I planned to make no spare parts, a risky practice that leaves no room for error.

Roughing out the pieces—Each door was built to the exact dimensions of its existing opening, then trimmed to fit. I ripped the door parts on a table saw and cut them to length on a radial-arm saw. When determining the lengths of the rails and mullions, I allowed for the depth of the coping profile—in this case ¹⁷⁄₃₂ in. The mullions were left long until the stiles and rails were machined and assembled dry. That allowed me to lay out the exact lengths of the mullions before cutting them.

Doweling—Dowels replace traditional mortise-and-tenon joints and are a critical adjunct to cope-and-stick joinery. Doweling typically precedes coping and sticking because layout and drilling are most easily accomplished before the frames are molded. I used 1-in. dia.

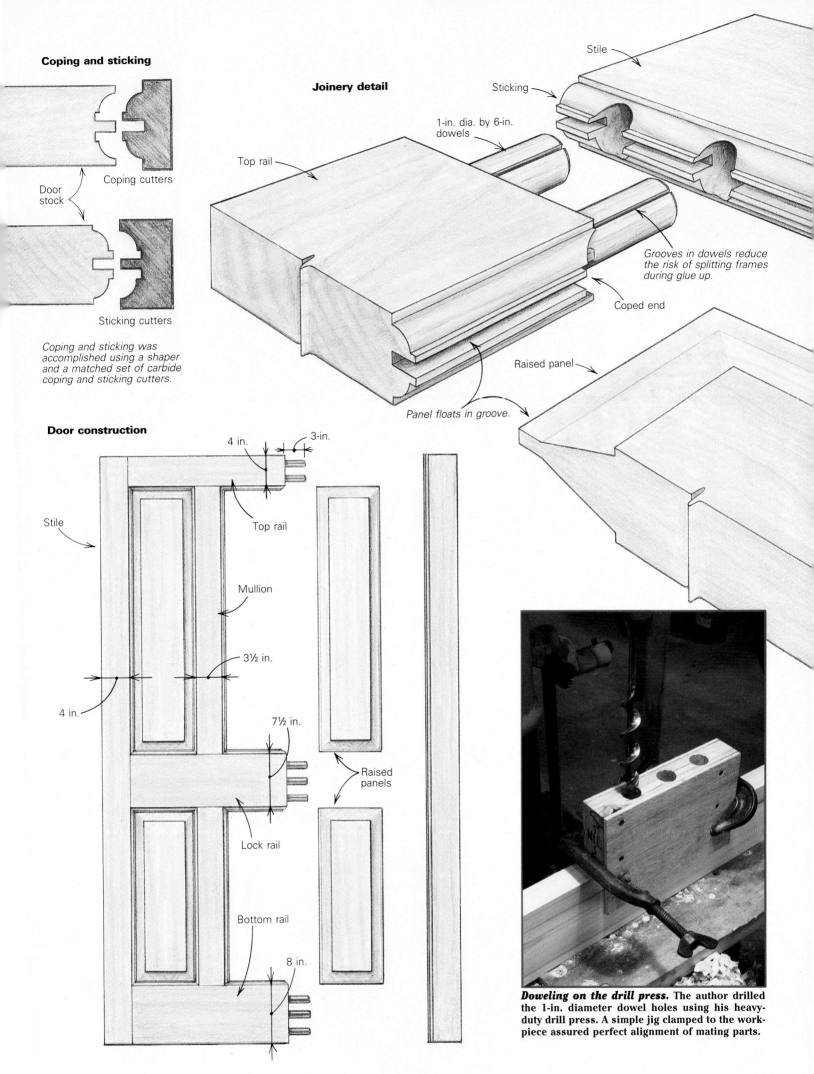

Coping and sticking

Coping cutters

Door stock

Sticking cutters

Coping and sticking was accomplished using a shaper and a matched set of carbide coping and sticking cutters.

Joinery detail

Top rail

1-in. dia. by 6-in. dowels

Stile

Sticking

Grooves in dowels reduce the risk of splitting frames during glue up.

Coped end

Raised panel

Panel floats in groove.

Door construction

Stile

4 in.

3-in.

Top rail

Mullion

3½ in.

4 in.

7½ in.

Raised panels

Lock rail

Bottom rail

8 in.

Doweling on the drill press. The author drilled the 1-in. diameter dowel holes using his heavy-duty drill press. A simple jig clamped to the workpiece assured perfect alignment of mating parts.

by 6-in. long dowels: two at each end of the top rails and three at each end of the bottom and middle rails. Spaced 2¼-in. o. c., the dowels are held at least ⅝ in. away from the edges of the rails to allow milling and trimming of the doors without cutting into the dowels. The cope-and-stick joinery alone suffices for anchoring the ends of the mullions to the rails.

I drilled 3⅛-in. deep dowel holes using a shop-made guide in tandem with a heavy-duty drill press to ensure precise alignment of the mating parts (photo, p. 81). The dowels were cut from 3-ft. lengths of standard 1-in. dowel stock, but not before I ripped on the table saw a pair of ⅛-in. deep saw kerfs 180° apart. These grooves would allow glue to squeeze out the bottoms of the dowel holes during glue up, relieving pressure and reducing the risk of splitting the frames. When cutting the grooves, I guided the dowel stock by feeding it between the standard rip fence and an auxiliary fence, which were installed about 1 in. apart on opposite sides of the sawblade. Because the depth of cut was a meager ⅛ in., I felt safe using this method for cutting the saw kerfs. I would devise a different method for making deeper cuts in round stock, however, because the stock *could* rotate while being fed through the saw and kick back.

Once I cut the dowels to length, I drove each one through a 1-in. dia. hole that I drilled through a ⅜-in. thick steel plate, ensuring dimensional consistency (the dowels initially measured slightly more than 1 in. in diameter and were slightly oval). Finally, using a stationary disk sander, I tapered the ends of the dowels to ease installation. The completed dowels were then set aside until glue up.

Coping and sticking—There are a variety of carbide cope-and-stick shaper cutters on the market. Most come in matched sets: a sticking set and a coping set. The sticking set typically consists of two molding cutters separated by a straight cutter; the straight cutter mills a groove for holding door panels. The coping set consists of two molding cutters ground to an exact inverse of the sticking cutters and includes a spacer instead of a straight cutter (detail drawing previous page). The spacer leaves a tongue on the ends of the rails and mullions that engages the panel groove produced by the sticking cutters.

My carbide cutters, which are made by Freud (218 Feld Ave., High Point, N. C. 27264; 800-334-4107), come with shims for fine-tuning the fit between the coped tongue and the panel groove, but I've never had to use them.

Before coping and sticking the door parts, I experimented on wood scraps, adjusting the setups where necessary. When cutting the door parts, I cut all the copes first, then the sticks. That way, any splintering produced by the coping cutters at the trailing edge of the stock (common when cutting wood across the grain) would be removed by the sticking cut. I also limited tearout during coping by backing up the work with a wood scrap. When sticking the stiles, I used infeed and outfeed rollers to help support the ends, and hold-downs (steel clips) to keep the work flat against the table.

Panel raising—Panel dimensions can be calculated from plan drawings, but that method invites mistakes. I prefer to knock the door frames together dry, without dowels, and to measure for the dimensions of the panels directly from the frames. I measure to the bottoms of the panel grooves, then subtract about ⅛ in. per foot of panel width to allow for seasonal expansion of the wood (for more on the seasonal movement of solid wood, see *FHB* #69, p. 54).

The panels were raised on the shaper using a carbide raised-panel cutter. Prior to shaping, however, I roughed out the bevel cuts on the table saw to remove the bulk of the material. This allowed me to run the finish cut on the shaper in one easy pass, reducing both machine time and wear on the panel-raising cutter. The panels are raised on both sides, so I adjusted the finish cut to leave the correct tongue thickness (about ¼ in.) where the panel edge is captured in the door frame.

Safety on the shaper—For convenience with the long, heavy door panels, I ran the shaper cutter submerged (rotating clockwise below the stock). This is opposite the usual procedure, in which the cutter rotates counterclockwise *above* the stock. My method has two advantages: the stock shields the cutter from fingers, and if the panel is inadvertently bumped on the shaper table, the only result is a crown that can be removed quickly with a second pass. The only drawback to this method is the mass of chips that can accumulate around the cutter, often in sufficient volume to heat the table and bog down the motor. I stopped the shaper after raising each panel and blew out the waste with compressed air. This interruption can be annoying, but it leaves you with perfect panels and a full complement of fingers.

Gluing up—I assembled the doors with aliphatic-resin glue, which is sufficient for interior work. When assembling the four, four-panel doors, I began by gluing up sub-assemblies of three rails and two muntins. Then I slid the panels into place, brushed glue into the dowel holes, drove the dowels into the stiles and, finally, drove the stiles onto the rails. Using a pair of pipe clamps across each rail, one over and one under, I drew the joints up tight and true. On the two-panel closet door, I simply assembled a stile and the three rails, inserted the raised panels, drove on the opposing stile and clamped the door tight.

Because of the long bearing surface at the ends of the wide rails, the doors drew up square without effort. If they hadn't, I would have corrected them by skewing the clamps off square to pull the doors into alignment and checking for square by measuring the diagonals (diagonals that are equal in length indicate a square door). The long dowels kept the joints relatively flat, and any deviations were straightened out by tweaking the clamping pressure. Shortly after assembly, I moved all the panels within their grooves to ensure that they weren't inadvertently caught by squeezed-out glue inside the joints.

Coping in the round—To add light at the end of the hall, I fitted the master-bedroom door with a pair of round-top, leaded-glass panels that I made in the shop (photos facing page). The semicircular panel heads echo the vaulted ceiling above, and the colored glass casts a soft, almost ecclesiastical light at the dark end of the hall without compromising privacy in the bedroom.

The door was assembled like the others but without raised panels in the upper two openings. After assembly, I glued up four simple filler blocks, coped them on the shaper and glued them into the upper corners of the panel openings. Next, I laid out curved cutlines on the blocks using a pencil compass and rough-cut the curves with a jigsaw. That done, I molded the rough-cut curves on the shaper, guiding the cut with a ball bearing stacked on top of the sticking cutters and a plywood template screwed to the door.

To permit the installation of the leaded-glass panels, I removed the molded edge on the back side of the door using a router chucked with a straight bit and guided by another ¼-in. plywood template clamped to the door. The glass panels are centered in the openings with several daubs of clear silicone caulk and locked in place with removable molded wood stops.

I made the straight stops by sticking the jointed edge of a poplar board, then ripping off the molded stop on the table saw. To make the curved stops, I began by gluing up a block of scrap poplar, bandsawing a radius in it and sticking the radiused edges on the shaper using the same bearing/template system used for sticking the curved edges in the door. The curved stops were then cut from the blocks on the bandsaw and dressed to a precise fit on their convex sides using the disk sander.

With the glue ups completed, it was time to clean up the doors and smooth the joints. Some doormakers accomplish this with a belt sander, but I prefer to use a sharp bench plane.

Making casings—Before hanging the doors, we made and installed poplar casings to match. The casings (photo, p. 80) are designed to echo the classic colonial detailing in the recently renovated living room adjoining the hall. I milled a double-ogee pattern into the 4-in. wide side casings using a single-knife molding head on the table saw. This molding head is an old tool that looks primitive, but it can produce a very fine finish. Because there is only one knife to grind for any particular segment of a profile, custom patterns can be set up with little difficulty. I achieved the desired profile by making two

1. Filling the corners. To accommodate the arched, leaded-glass panels in the master-bedroom door, the author glued up four filler blocks, coped them on a shaper and glued the blocks into the panel openings.

2. Shaping round the bend. Next, Beals rough-cut the arches using a jigsaw and molded the radiused edges on the shaper, guiding the cut with a ball bearing riding against a plywood template screwed to the door.

3. Routing recesses. To allow installation of the panels, the author then removed the sticking on one side of the door using a router guided by a second plywood template screwed to the door.

4. Curved stops. Curved stops were made by bandsawing their inside radius in a poplar scrap, molding the radiused edge on the shaper (guided by a template), bandsawing the outside radius and smoothing with a disk sander.

passes with a concave cutter and two passes with a convex cutter.

The side casings are cut square at the top and surmounted by an architrave head casing, sometimes called a cabinet head. Architrave head casings range from straight and simple to ornate arched and gabled designs, all of which represent the entablature of classical architecture. In contemporary work, where mitered casings prevail, even a simple architrave head casing has an air of classic elegance.

I fitted four of the five new doors with a straight head casing. This casing consists of ¾-in. by 5¼-in. pine, with a simple shaper-cut bead applied to the bottom edges and a standard 1¼-in. pine bed molding applied across the tops. I returned the ends of the beads using a sharp block plane and sandpaper. The bed moldings have full mitered returns, glued and bradded in place.

The head casing on the leaded-glass door takes the classic shape of a broken pediment and urn, a detail that flows gracefully below the arch of the curved ceiling directly above (photo right). A 1x backplate is bandsawn to the pattern and, like the straight head cases, has a bead applied to the bottom and bed moldings applied and returned on the top.

A glass-top door. The master-bedroom door is fitted with a two leaded-glass top panels, which help illuminate the hall without sacrificing privacy in the bedroom.

Next, I turned the urn on the lathe in a simple, classic profile and then bisected it on the bandsaw. The head case is coped against the closet head case on the left-hand side.

The urn was installed last. I simply put several daubs of silicone adhesive on the back and pressed it into place.

Hanging and finishing—The doors were trimmed to fit their openings with a hand plane and beveled at a 3° degree angle on the lock stiles so the doors would swing clear of the jambs. I cut hinge mortises in the new doors to match the existing jamb mortises, and then hung the new doors on the old hinges. Rather than reuse the old passage locksets, we installed new lever locksets.

After covering everything with an oil-based primer, we painted the hall ceiling a standard latex flat white, the walls a latex semigloss sky blue and the woodwork a semigloss ivory oil enamel. The transformation of the hall is very satisfying, from cheap and dark to a pleasing harmony of crisp, bright details. □

Joseph Beals is a designer and builder in Marshfield, Mass. Photos by author except where noted.

Pueblo Modern
With knife and brush, a furnituremaker applies Southwestern motifs to a houseful of doors

by Craig W. Murray

I was fortunate to grow up in a house full of light and color. My architect father and artist mother designed our house, and living in it infused all of their children with an appreciation of art and craft. I want my children to grow up in a similar atmosphere, so when I built our new home in the mountains of Albuquerque, one of my goals was to make its interior detailing unique and surprising, reflecting the timeless quality of the Southwest. I wanted to combine my New England upbringing (and my memory of simple frame-and-panel construction) with the design vernacular of the High Plains. We needed a houseful of doors, so beginning there I designed each without repeating myself. Cabinet doors were the next target.

I approached my door-building resolution by studying native pottery, basketry, weaving and embroidery. There is a rich heritage of such art and craft, and the bold geometric textile patterns and softly burnished ceramic shapes that characterize it reappear over and over. I learned that Indian, Hispanic and Anglo influences long ago combined to produce the familiar angular designs and earthy tones played against intense color that we recognize today as Southwestern.

As I dug into the architectural history of the region I found, not surprisingly, that the lineage of wooden doors in the Southwest doesn't stretch back very far. Hides and blankets served the purpose before milled lumber became widely available in the 19th century. Most of the local doors I did find had wood-strip appliqués nailed to the panels. A few had rudimentary carvings, with minimal use of color. That became my point of departure, as I began my door series using a knife to carve the panels and artist's oils to color them.

Notes on making the parts—I used native Ponderosa pine for door frames and door bucks (the thick wood jambs found in adobe buildings). For doors, I used a grade called D-select. It has some wild grain and tight knots in places and costs less than the clearer grades.

Painted black. **Rabbit and trout motifs from 15th-century Mimbres pottery inspired the closet doors in the children's bedrooms.**

But I think the variations add interest, and most of it is clear enough that I can easily pick out straight sticks for stiles. For the entry door I started with 12/4 stock. The remaining exterior doors and all interior doors have 8/4 frames, while closet doors have 6/4 frames. Most of my interior doors have Ponderosa pine panels, but I used cherry, ash and walnut panels for the exterior doors.

I rough-out the tenons with a radial-arm saw and a table saw. Then I cut away most of the unneeded stock with a bandsaw and clean up the tenon with a router and a straight bit.

A drill press serves nicely to rough-out the mortises, after which they can be cleaned up with a slick and a mortising chisel. I make the panels out of 4/4 stock and use a shaper to give them a coved or rabbeted edge. After dry-assembling the doors and panels to check the fit, they get a sanding with 220-grit before I start carving and coloring.

Layout and carving—I typically draw my design on heavy paper and use a razor knife to cut out the shapes. With the pattern on a door panel, I lightly draw the shapes in pencil on the panel, using the cutouts as a guide. I go over the pencil lines with a knife, making a cut about $1/32$ in. deep. The pencil lines

Murray reprises the diminished-stile and raised-panel design typically associated with New England detailing by combining bird's-eye maple panels within orange rails and stiles. The author first colored them with cadmium red, followed by a second pass with deep cadmium yellow.

After completing the carving of passage doors throughout the house, Murray turned his attention to the cabinetry. The dancing triangles in the rail were inspired by the embroidery of Hopi sashes, while the stacked V's in the top panel are reminiscent of Navajo rugs.

Murray's most recent cabinet combines the zigzag spindles typical of Southwestern stair and balcony railings into a floating panel. After roughing out the spindles on the bandsaw, he used a 1-in. belt sander to erase the saw marks.

Simple zigzag patterns in various widths can be used to good effect, especially when combined with complementary colors. To preserve the original lustre of the doors, Murray gives them periodic rubdowns with an oil finish/solvent.

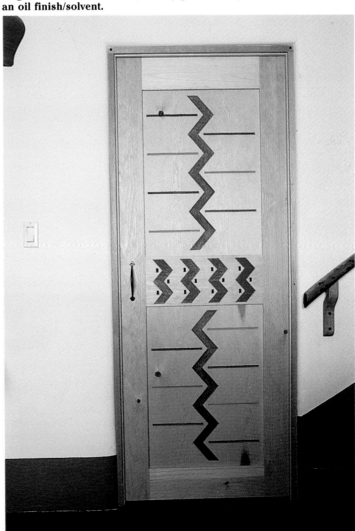

shouldn't show up on the finished door, so I go over the whole panel with 220-grit sandpaper after the first pass of the knife. This makes the panel ready for its first coat of oil and wax.

This is a critical step in the process, and one that I learned the hard way. Color applied to a raw incision in wood will wick into the pores—both down and sideways. This is fine if it goes where you want it, but if you want a crisp edge, it can be maddening to watch color bleed horizontally into the surrounding area, making fuzzy blotches where it should be clean. The secret to preventing color bleeding is to apply the oil and wax after lightly making the outline with the knife, sealing the top edges of the carved incisions.

In a well-ventilated room, I melt 3 to 4 oz. of beeswax (for a 1-qt. batch) in a pot over a hot plate. Then I add 3 parts boiled linseed oil and 1 part gum turpentine to the pot. It's important to use boiled linseed oil—unboiled won't dry. With a rag, I rub this mixture thoroughly over any part of the panel I don't want to color.

I can go to work with the knife as soon as I've applied the oil and wax mixture. I've tried chisels, parting tools and gouges to incise the lines in the panels, but nothing works as well for me as a carving knife (top photo, right). For most line work, I hold the knife at about a 45° angle, making a V-shaped groove about ¼ in. to ⅜ in. deep. I keep the knife surgically sharp with a black Arkansas stone.

People ask why I don't use a router for this operation. It would often take longer to set up the router guides than to carve the design by hand, for one thing. And part of the interest is the crisp look that carving imparts to the work.

Applying color—Using powdered pigments or the artist-grade oil colors that come ready-mixed in a tube, I brush and wipe color onto stiles, rails and panel carvings (center and bottom photos) in subtle combination or wild concoction. To prepare a color batch, I spoon a small amount of powdered pigment or a short squeeze of a tubed color into a cup. Artist's pigments are available at art-supply houses as a fine powder packaged in jars. The more familiar tubes of linseed oil-based colors work as well and come in a greater range of colors.

To the color I add a teaspoon or so of boiled linseed oil or natural Watco Danish Oil Finish (manufactured by Minwax Co., Inc., 15 Mercedes Dr., Montvale, N. J. 07645; 201-391-0253), a teaspoon of gum turpentine and a few drops of Grumbacker Oil Painting Medium III to speed drying time (available at art-supply stores). It's important to experiment with the colors to see how they look on the type of wood in question. I've got sample blocks of pine in dozens of colors lining one wall of my studio as a reference library.

A color's permanency will be listed on the label. Choose the ones with the highest rating. They will cost more than less permanent colors, but the difference is insignificant because a little goes a long way. Note that some colors have toxic ingredients in them. Those, too, will be listed on the label, and be sure to han-

After lightly scoring the pencil lines with a knife, Murray coats the panel with a mixture of turpentine, beeswax and linseed oil. During carving, the panel is held to the bench by the friction of a rubber carpet pad. Incisions are typically V-shaped in section.

Murray uses an artist's brush to apply the color to the carved grooves. Thinning the paint mixture helps it sink into the grain of the wood. The oil and wax primer atop the panel keeps color from bleeding beyond the boundaries of the design incisions.

Broad surfaces, such as the edges of panels with coved edges, are best colored with a rag because vigorous rubbing drives the color into the grain, evening out the tone. While the paint is still wet, other colors can be worked into the grain to change the original color.

dle the material with care. For instance, whenever I use paints with cinnabar or manganese in them, I wear rubber gloves and a mask rated for hazardous vapors. All paints should be applied in a well-ventilated space.

The paint mixture should be thin enough to penetrate the wood grain quickly, and the color should be intense but not opaque. Colors can be applied over one another, so if one isn't looking the way I expected, I'll modify it with an overlay of another color. The colors are translucent, and underlying grain patterns of early and late wood in the pine stand out against the vivid colors and designs on them.

I let the colors dry overnight before gluing up the door. Then I finish the assembled door with several coats of the heated beeswax/linseed oil mixture, which is rubbed on and buffed off, adding lustre to the color.

Hardware and maintenance—I hang the doors with simple square-corner butt hinges. Although they are available in black, I bought primed hinges and painted them myself to save about 40% of the cost. I mortised the hinge gains in the doors and bucks with a template-guided router, and squared the rounded corners of the gains with a chisel.

Brassy doorknobs wouldn't look right on these doors. Fortunately, a shop right up the road in Santa Fe specializes in hand-forged ironwork, and I selected their pulls and latches for the doors (Tom Joyce Architectural Blacksmithing, Rte. 9, Box 73J, Santa Fe, N. M. 87505; 505-983-0880).

The linseed oil/wax finish gives the doors a rich sheen, but they do collect fingerprints and other smudges. To keep them clean, I give them a periodic rubdown with Watco Satin Maintenance Oil. It has a solvent in it that cuts through the grime, and an oil that preserves the doors' original lustre. □

Craig W. Murray designs and builds nontraditional furniture, doors and light fixtures in Cedar Crest, N. M. Photos by Charles Miller.

Suggested reading
Following is a list of books that I found to be good sources for Southwestern art and craft.
- *Navajo Textiles: The William Randolph Hearst Collection* by Nancy J. Blomberg. University of Arizona Press, Tucson, Ariz., 1988. $45, hardcover; 257 pp.
- *Mimbres Painted Pottery* by J. J. Brody. From the School of American Research Southwest Indian Arts Series. University of N. M. Press, Albuquerque, N. M., 1977. $27.50, softcover; 276 pp.
- *The Navajo Weaving Tradition: 1650 to the Present* by Alice Kaufman and Christopher Selser. E. P. Dutton, New York, N. Y., 1985. $24.95, softcover; 150 pp.
- *New Mexican Furniture: 1600 to 1940; the Origins, Survival and Revival of Furniture Making in the Hispanic Southwest* by Lonn Taylor and Dessa Bokides. Museum of N. M. Press, Santa Fe, N. M., 1987. $49.95, hardcover; $39.95, softcover; 336 pp.

Pocket Doors
Should you buy a kit or build your own?

by Kevin Ireton

When I asked one architect about pocket doors—doors that slide into a wall, rather than swing on hinges—he said he avoided them like the plague. Another told me he used them only as a last resort. One builder simply said, "They're stupid. Don't use them." Modern-day pocket doors have a reputation for creating flimsy walls on either side of the pocket, for using lightweight hardware that's easily misaligned and rollers that jump off the track, and for needing repairs that are impossible without tearing open the wall.

But blaming pocket doors for these problems is like blaming a circular saw for not cutting straight. Having installed a few pocket doors myself, and having recently spoken to builders all over the country about pocket doors, I've learned that, properly installed, some commercially available frame kits work just fine. I also discovered a great system for building your own pocket-door frames.

Sliding versus swinging—To swing 180°, a standard 2-ft. 8-in. door needs over 10 sq. ft. of clear floor space. A typical house—three bedrooms, two baths—might have nine or ten swinging doors whose collective door swings lay claim to 100 sq. ft. Half of that figure represents the space directly in front of the door, which has to remain clear anyway. But the other half is usable space—50 sq. ft. of it—that could be reclaimed through the use of pocket doors.

Pocket doors are commonly used for two specific reasons: either because the space and traffic pattern demand it, such as in a half-bath off a hallway where it would be awkward to have a swinging door; or because you want the option of occasionally closing off a room without having to sacrifice space in return. An example of the latter would be pocket doors used between kitchens and dining rooms. Such doors are going to be in the pocket 90% of the time, but when you want to hide the dirty dishes after Thanksgiving dinner, you can pull them shut.

One advantage of pocket doors is that the two sides can be painted different colors, or even milled differently, to match the rooms they face. A swinging door, on the other hand, shows both sides to the room it swings into—one side when the door is open, the other side when it's closed.

Structurally, pocket doors have the disadvantage of requiring a header that's twice as long as that of a swinging door. The problem is magnified with converging pocket doors—a pair of pocket doors that slide toward each other. For example, if you have a 6-ft. opening between two rooms and you're debating double swinging doors versus converging pocket doors, the latter will require a header that's nearly 13 ft. long. Another disadvantage of pocket doors is that if they're used in a 2x4 wall, you don't have room for an electrical outlet or switch in the area of the pocket. None of these problems is insurmountable, though, and if you don't have room to swing a door, sliding it into a wall may be your best option.

Pocket-door frame kits—Two types of pocket-door frames are available off the shelf, for use in 2x4 walls (though they can be furred-out for thicker walls). One type comes sized for specific doors and includes two preassembled wall sections and a header/track assembly. The wall sections are ladder-like affairs made of 1x stock, with two vertical pieces connected by three or four horizontals.

Nobody I've talked to likes these kinds of frames. The tracks are usually light-gauge steel with single-rollers on the door hangers, and they're rated only for 50-lb. or 80-lb. doors. (Although most standard interior doors weigh less than 50 lb., the weight rating offers some indication of how easily the door moves on the rollers.) The lumber used in these units is of poor quality, seldom straight, and the installed wall sections are flimsy. Even one of the manufacturers I talked to (who also makes the other kind of door frame) told me it wasn't a good frame.

The better frames are "universal" pocket-door frames, available now from a number of

Universal-type pocket-door frame. The header/track assembly (top left) is nailed to the end studs, not to the framing header, which allows the header to sag a bit without bending the track. Wrapped with steel on three sides, the pocket-door studs are screwed or nailed to the sides of the header/track assembly at the top (top right), and at the bottom, slip over prongs on a metal bracket nailed to the floor (bottom right). A rubber bumper, attached inside the pocket (bottom left), limits the travel of the door and prevents it from banging into the end stud.

companies (see chart, facing page). They cost around $40 or $50 (not including the door) and can accommodate most doors that are 2 ft. to 3 ft. wide and 6-ft. 8 in. high (some companies offer frames or extension hardware for doors up to 5 ft. wide and 7 ft. high). The frame kit includes a header/track assembly, two door hangers and four 1x2 studs. The studs are wrapped with steel on three sides to stiffen the wall and prevent nails and screws from penetrating into the pocket (top right photo, previous page). The tops of the studs should be nailed into the side of the header/track assembly, but the bottoms slip over the prongs of a metal bracket nailed to the floor (photos previous page, top and bottom right), which lets the floor sag a bit without the studs pulling the track down and bending it.

The door hangers supplied with these frames have at least three nylon rollers (some have four), and the tracks are shaped so that the hangers have to be slipped over the end of the tracks, which makes them virtually "jump-proof." Most of the universal frames now come with a two-piece hanger. The roller unit is one piece, and it slips into the track with a threaded stud that hangs down. The other piece is a bracket that you attach to the top of the door.

When you attach the brackets, be sure to place them far enough from the end of the door that both screws go into the top rail, not into the end grain of the stile. After the brackets are attached, the door is lifted into place, the bracket engages the stud and is usually locked by a little plastic gate that you swing shut. The door can be adjusted by screwing the stud in or out of the roller housing with a special wrench included with the kits.

One of the criticisms of pocket doors is that you have to tear out the wall if something goes wrong, but some of the manufacturers (Johnson, Sterling and Acme, for instance) have designed their tracks so that you can remove them for repair or replacement without disturbing the wall. After taking down the door, you remove the mounting screws in the section of track above the doorway. The remaining screws, inaccessible in the pocket, sit in keyhole-shaped slots; simply pull the track toward you a fraction and the track will slip over the screws. Before installing the track, check that these screws are not too tight or you'll never get the track out later.

Most of the universal kits are rated for 100-lb. or 125-lb. doors. Some companies offer heavier duty hardware if you need it. Acme has a universal frame rated for 250-lb. doors.

The Woodmeister Corp. builds pocket-door frames as a unit in the shop and ships them to the job site squared and braced. Above, Jeff Ham plumbs the strike-side of the pocket-door frame. The plywood plate running across the doorway will be cut out after the frame installation.

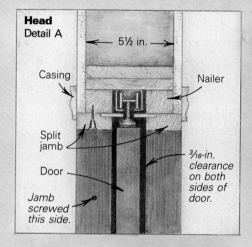

Head Detail A

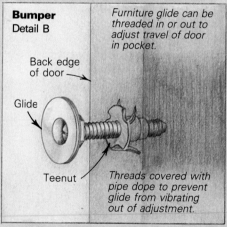

Bumper Detail B

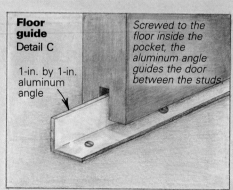

Floor guide Detail C

And Johnson Products has a set of optional ball-bearing hangers rated for 200-lb. doors. With a 125-lb. door and Johnson's standard rollers, the folks at Johnson say it takes between 6 lb. and 7 lb. of pull to slide the door out of the pocket. With a 200-lb. door and Johnson's ball-bearing hangers, it only takes 3 lb. of pull.

When I asked if the ball-bearing hangers would hold up any better, the folks at Johnson said that they didn't think so. They've cycle-tested their standard rollers using a 150-lb. door, and after 100,000 trips in and out of the pocket, all they got for their trouble was a black stripe around the white nylon wheel.

Frame installation—Installing a universal pocket-door frame is not difficult, but you have to be careful. The trouble is that pocket doors must be installed while you're framing the house, before drywall, and in a mood to work quickly rather than carefully. The manufacturer's instructions explain the basic installation, but here are some things to keep in mind.

The track should be nailed between the trimmer studs on either side of the rough opening. The track is not attached to the framing header, so there should be a space between them (top left photo, p. 87). This allows the header to sag a bit without bending the track.

Before you install the studs, site down their length. Three out of four studs in the last frame I bought had a ¼-in. bow in them. You can straighten them with your hands, which indicates how easily they can be knocked out of alignment. The studs should be carefully

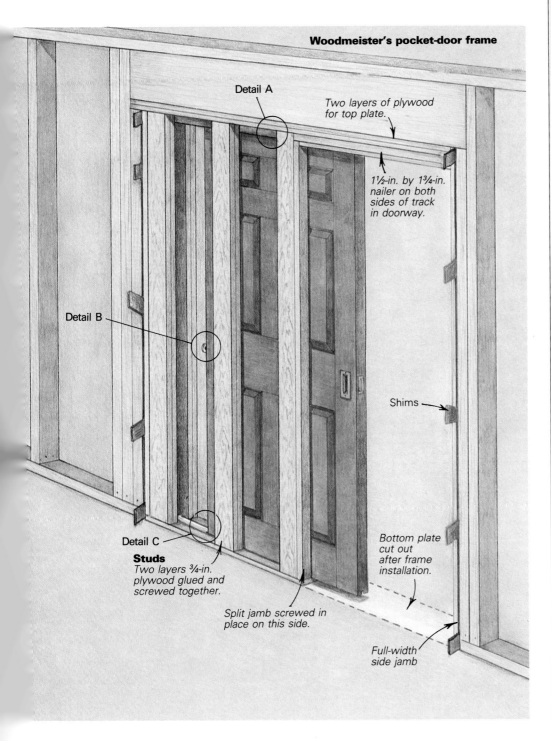

Woodmeister's pocket-door frame

Manufacturers of pocket-door frame kits and hardware

	Pulls and latches	Tracks and rollers	Frame kits
Acme General Corp. Div. of the Stanley Works 300 East Arrow Highway San Dimas, Calif. 91773	•	•	•
Baldwin Hardware Corp. 841 E. Wyomissing Blvd. Reading, Pa. 19611	•		
G-U Hardware, Inc. 11761 Rock Landing Dr. Suite M6 Newport News, Va. 23606	•		
Grant Hardware Co. High St. West Nyack, N. Y. 10994-0600	•	•	
Hafele America Co. 3901 Cheyenne Dr. P.O. Box 4000 Archdale, N. C. 27263	•	•	
Iseo Locks Inc. 260 Lambert St. Suite K Oxnard, Calif. 93030	•		
L. E. Johnson Products, Inc. 2100 Sterling Ave. P.O. Box 1126 Elkhart, Ind. 46515	•	•	•
Lawrence Brothers, Inc. 2 First Ave. P. O. Box 538 Sterling, Il. 61081	•	•	•
Merit Metal Products Corp. 242 Valley Rd. Warrington, Pa. 18976			•
National Manufacturing Co. 1 First Ave. Sterling, Il. 61081	•	•	•
Quality Hardware Manufacturing Co., Inc. 12705 S. Daphne Ave. Hawthorne, Calif. 90251			•
John Sterling Corp. 11600 Sterling Parkway Box 469 Richmond, Il. 60071-0469	•	•	•
Wing Industries, Inc. 1199 Plano Rd. Suite 110 Dallas, Tex. 75238			•

plumbed in both directions, especially the two at the end of the pocket that will support the split jamb. It's a pain in the neck to shim the split jamb to make it plumb.

After you've installed the frame, insert a couple of temporary spacers (1x stock, cut to the width of the pocket) horizontally in the pocket to prevent the studs from bowing inward while the drywall is hung. Even with the spacers in place, though, whoever is hanging the drywall should be careful not to bear down so hard on the screwgun as to knock the studs out of alignment. Shorter screws should also be used.

Be sure the full-width side jamb that the door closes against is plumb. If this piece and the split jamb opposite it are not both plumb, then you can't adjust the door so that it both closes properly and rests flush with the split jamb when open.

To allow access to the hangers (in case you have to remove the door later), you'll need to screw one side of the head jamb in place rather than to nail it. This means you should put the side jambs up first; otherwise they'll trap the head jamb. Also, remember not to nail the casing into the head jamb that's screwed in place.

Make sure that the door you use is straight and flat, and that it's sealed on all sides, including the bottom. If the door warps, it will hit the studs in the pocket and won't open or close. Remember, too, that more than one pocket door has gotten scratched or nailed in place by a carpenter who forgot (or didn't know) that he was nailing into a thin wall. It won't hurt to remind subcontractors about the pocket door, either.

Building your own—If you have to install a pocket door in a 2x4 wall, you might as well use a kit. Although I've spoken to some builders and architects who think otherwise, most agree that the kits work pretty well. On the other hand, if you can afford to beef up the finished wall thickness to 6½ in. (2x6 framing), you're probably better off buying the hardware and building your own frame. This is especially true if you're using bigger than average doors. Besides letting you construct stiffer walls on either side of the pocket, a 2x6 wall gives you enough room to install shallow electrical boxes.

The best system I've seen for building your own pocket-door frame is the one worked out by the people at the Woodmeister Corp. (drawing previous page), an architectural woodworking company in Worcester, Mass. They start with heavy-duty tracks and rollers (Woodmeister uses Lawrence hardware, but Grant and Acme also make some pretty rugged stuff). Then they build the frame using ¾-in. cabinet-grade veneer-core plywood (poplar or birch) rather than solid lumber, because it's more stable.

The entire frame is assembled in the shop. It's built as a unit to fit the pocket-door's rough opening (photo, p. 88). The top and bottom plates run the full length of the rough opening, with a 5½-in. wide stud at each end. The bottom plate, which is continuous across the door opening, is cut out later. The studs are made up of two layers of plywood, each 3½ in. wide. Before the layers are glued and nailed together, the fabricator sights their length and orients the pieces so that any bow in one is countered by the bow in the other.

In addition to the overhead track, Woodmeister uses a continuous floor guide inside the pocket that centers the door and prevents it from hitting the studs. A piece of 1-in. by 1-in. aluminum angle, running the length of the pocket and extending about ¼ in. past the last stud, is screwed to the bottom plate (bottom drawing, p. 88). A corresponding groove is routed in the bottom edge of the door to receive the angle. After assembling the frame, the fabricator squares it up and screws a full-length diagonal brace across it before shipping it to the job site.

Installation is straightforward. As with the off-the-shelf frames, the top is not attached to the header of the rough opening, and the important points are that the track is level and the studs plumb in both directions. Woodmeister's installers shim the frames and screw them in place to avoid the possibility

The edge pull is used to get the door out of the pocket, and the flush pulls, installed on both sides of the door, are used to open the door (photo above). But with this setup, there is no way to latch or lock the door. The unit shown in the top photo combines face pulls, edge pull and privacy lock all in one. To install it, you simply cut a notch in the edge of the door; no mortising is required.

of repeated hammer blows knocking the frame out of alignment.

The standard frame kits supply you with a little rubber bumper to nail on the end stud in the pocket to cushion the blow from the door hitting it (bottom left photo, p. 87). Woodmeister installs a teenut and an adjustable plastic floor glide (like the kind used on the legs of office furniture) in the back edge of the door (middle drawing, p. 88). This cushions the blow of the door but also allows adjustment of the travel of the door into the pocket. Before installing the floor glide, pipe dope is smeared on the threads of the glide to keep it from vibrating loose over time.

When installing the split jamb, the folks at Woodmeister screw one side of both the side and head pieces to the plywood studs. This makes removal of the door easier.

Most people don't use any kind of door stops on a pocket door. The leading edge of the door simply butts into the jamb. Woodmeister has tried cutting a ¼-in. deep rabbet into the full-width side jamb to receive the door edge. But while this looks nicer when the door is closed, it doesn't look as good with the door open. Also, the use of a rabbet or stops can lead to trouble if the door warps at some point down the road.

Pulls and latches—During the Victorian period—probably the heyday of the pocket door—some wonderful decorative hardware was available for pocket doors, including ornate recessed pulls for the face of the door and great locksets with edge pulls that popped out when you pushed a button. Unfortunately, no one that I know of is reproducing them, so you'll have to shop the salvage yards if that's what you're after.

Here's an overview of today's options. You can use an edge pull in the edge of the door (a nice one is available from H. B. Ives, A Harrow Co., P. O. Box 1887, New Haven, Conn. 06508) and a pair of flush pulls (available from various companies) on both sides of the door (bottom left photo). Or you can get a lockset that includes flush pulls, edge pull and a privacy latch in one unit (top left photo). The ones I've seen were made by Quality Hardware Manufacturing Co. (see chart on p. 89 for address). There's no mortising with this unit. Instead, you cut a 2¼-in. by 1¾-in. notch in the edge of the door. Such a big bite, though, could affect the integrity of some doors, especially hollow-core doors, and perhaps lead to warping. In any case, this type of lockset will likely void any warranty on the door.

Some of the companies that make frames (Lawrence and Johnson, for instance) sell latches and face pulls that fit the holes for standard swinging-door locksets. If you want real security, Baldwin, Hafele, Merit and Iseo (see chart for addresses) make case-style locksets available with key cylinders. □

Kevin Ireton is an associate editor of Fine Homebuilding.

About old pocket doors
by James Boorstein

I cannot put a date on the earliest use of pocket doors in this country, but I know they were used in the 18th century. As American architecture evolved beyond its rustic and purely functional roots and grew increasingly grand, large pocket doors became quite common. Possibly it was an easy way for some builders to deal with huge doors without having to use massive hinges and heavy framing. Pocket doors were used to separate the more public rooms of the house—the parlor, library and dining room—from each other. Rarely were they used on the upper floors to separate bedrooms.

Most of the early domestic pocket doors had wheels on the bottom and rode in a track on the floor. Not until the middle of the 19th century were overhead tracks and rollers available. The switch to an overhead system was probably made as hardware technology advanced. The overhead track was out of sight and was less susceptible to problems resulting from dirt in the track and from the floor settling.

The older, more traditional pocket doors were almost always used in pairs (photo below). Each door rolls on two wheels that are either solid or spoked and 4 in. to 5 in. in diameter. Most are seated in a permanent housing called a "sheave," which looks somewhat like a window-sash pulley. Many of the wheels, especially those found in the southern U. S., can be very decorative, even though they aren't seen. Some of the sheaves are just long enough to house the wheel itself, other sheaves are two or three times the diameter of the wheel with a horizontal slot that allows the axle of the wheel to move forward or backward as the wheel rolls and the door moves. I'm not sure of the exact function of this. It certainly reduces the wear of the axle on the housing and perhaps changes the balance of the door, making it easier to move.

The sheave is mortised into the lower rail of the door (drawing below right), shimmed and fastened with wood screws to allow the door to sit level. Only about ¼ in. of the wheel protrudes below the edge of the door. The edge of the wheel is grooved to fit over a ridge in the metal track, which is often bronze or brass, and sometimes steel. The floor track is often surface-mounted, but is sometimes recessed into the finished floor. In many older homes the floor track appears to be raised. This is usually a result of the floor around the track being sanded away over the years. Lowering the track would affect the operation of the door.

To keep the door vertical, two hardwood pegs (often oak) are mortised into the top rail of each door. Even on the finest doors these pegs are fairly crude. They protrude 2 or 3 in. up into a wooden track. The track has two recesses. The top edge of the door extends into the first recess, and the hardwood pegs ride in the second. There must be adequate clearance above both so that the doors can be lifted up and over the floor track for installation and removal. Occasionally the wooden upper tracks were constructed in such a way that they float and actually lift up as the door is being lifted.

Converging pocket doors have a center stop, generally a sturdy piece of cast hardware screwed to the upper track and shaped to receive the leading edge of both doors. This must be removed to take the doors out. The leading edges on a pair of converging pocket doors were often milled to fit into each other, like a shallow tongue and groove.

There is also a stop, or bumper, on the trailing edge of the door so that the door will come to rest at the proper place when it is slid into the pocket. This stop, which is never seen, is often a very crude block of wood attached to the door at approximately the height of the pulls.

It is common to see traditional round door knobs on pocket doors; the stop keeps these doors 4 in. to 6 in. out of the pocket or just less than the width of the rail. Usually the knobs are fixed and act only as pulls to open and close the door. Locks and latches on the old doors vary, but they all work on the same principle: a curved metal bolt arcs out of the lockset in one door and down into a receiving plate on the other door or on the wall. These locks are often operated by a short, decorative key that works from either side of the door. On doors that rest fully recessed into the pocket when open, flush with the jamb, I've seen an array of ingenious pulls and spring-loaded pop-out handles that are flush until called into action.

Making repairs—Occasionally old door pockets lie hidden behind contemporary walls. Often these hidden pockets still house their rolling doors, but getting them out can be difficult. If they have been hidden through several renovations, they may have electrical cables or plumbing lines run right through them. More common problems include warped studs in the pocket, or a warped door. The door could be off its tracks, it could have been inadvertently screwed or nailed in place, the track could have worked loose or there could be any number of other problems. A screw eye and a loop of rope attached at the top and bottom of the door will serve as a temporary handle to coax a stubborn door out of its pocket.

Once the door is out, get a couple of droplights or electric clip-on lights and shine them in the pocket (a flashlight is no substitute). By carefully examining the pocket and employing some common sense, you will probably be able to figure out and correct any problems.

Restoring the operation of a pocket door should take place after any necessary structural repair in the area has been accomplished. But it should be done before any final plastering, as you may need to make holes in the wall to make adjustments inside the pocket. If for some reason the walls can't be disturbed (if there's wallpaper or a mural, for instance, or if the wall is otherwise historically significant), you can remove material from behind the baseboard and gain access to the pocket and the track without damaging the walls above.

Sometimes the door itself needs basic structural repair. The mortises for the wheels on the lower rail of the door frequently weaken the door. A partial or complete patch, with a new and relocated mortise, will solve the problem. The tracks should be straightened and shimmed level. On older doors the wheel bearings may be worn out. The simplest way to repair this is to pad out the mortises and install replacement sheaves (available from Grant Hardware, see chart on p. 89 for address). Usually, they will work on the existing track. On historic pocket doors I have had the worn hardware remachined, which worked well. While I retained the original hardware, however, I also preserved the original slightly stiff operation of the door. A compromise would be to use a better material for the bearing.

James Boorstein is a partner in Traditional Line, *a restoration company in New York City.*

Nine ft. tall and 3 in. thick, these converging pocket doors reside in the Dakota, one of New York City's most famous apartment buildings.

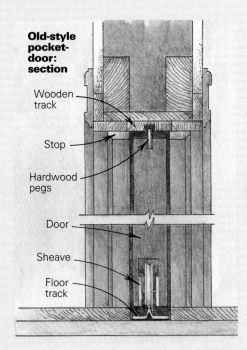

Old-style pocket-door: section
- Wooden track
- Stop
- Hardwood pegs
- Door
- Sheave
- Floor track

Curved Doors

A design with a circular closet in the center of the house calls for some special techniques

by Thomas Duffy

Several years ago, I was hired to redesign a small, single-story house to accommodate a family of three. The bungalow had been chopped up into a number of little rooms, and the new owners wanted more convenience and openness. It was obvious from the start that storage was going to be a big problem in this 24-ft. by 28-ft. space, and I wanted to make use of spots that might otherwise be wasted.

I began by dividing the floor plan into quarters and drawing in corner cabinets, as shown in the drawing, below right. Then I sketched in the bathroom and a short hall to get to it, a process that transformed the four corner cabinets into a single discrete structure at the center of the house. This pushed me out on a limb. Putting major storage dead center in a house plan isn't exactly a standard design solution. But it occurred to me to play with the idea of round storage space in place of the square closets or shelves, and I liked the result. With the rest of the house still laid out with straight walls, curves at the center would give it a sense of depth and spaciousness that might be surprising and delightful in such a small space.

Getting down to design details, I decided that the closet would be 7 ft. in diameter, containing kitchen storage, a linen closet and a bedroom closet, a sliding door pocket and a bookcase. A short section of curved wall in the bathroom hall would be the arc of a 14-ft. dia. circle swung from the same center. I'd recently found a quantity of butternut at a good price. I had never worked with this wood, and was eager to give it a try. My clients liked the way it looked, so we decided that all of the woodwork in the house would be butternut. As it turned out, the wood worked beautifully, and hand tools left it with a lovely, bright surface.

My crew's first job was to remove the existing partitions. Then, using trammel points, we traced the curve for the rough frame onto 2x12s. We bandsawed our top and bottom plates from this stock, and installed the top plate first, dropping plumb lines to locate the position of the bottom plate. We toenailed in the radial 2x4 studs of the closet's exterior walls, and installed door headers and cripples.

The door pocket was framed up out of 2x2s. Because of the curving surfaces, we decided to use plaster instead of drywall, so once the framing was done, we installed plaster grounds. These are 5-in. wide strips of wood that are nailed around the rough openings to be flush with the finished plaster surface. They look somewhat like casings, but are meant to act as guides or screeds while the wall is being plastered. They were removed and replaced with the actual casing after the plasterers and painters had finished their work.

Next, I installed the 4x4 center post, toenailing it to the subfloor below and to a nailer between the joists above. If the center post were taken out of the closet today, it might look like a crisp piece of sculpture, with all of the mortises, rabbets, notches and chamfers that were cut into it to accommodate the shelves, partitions and nailers.

I used interlocking joinery for two reasons. The first is philosophical: I like to make a piece of wood do as much as it can. I'd rather cut mortises than apply more wood to do the job. The second reason is practical: Unglued interlocking joints—dovetails, for example—can be taken apart and put back together again if any adjustments have to be made as the project unfolds.

I mortised horizontal shelf supports of 5/4 pine into the post, and toenailed their other ends to framing members. Then I nailed up vertical tongue-and-groove pine partitions between storage areas, and screwed cleats to the

The circular storage unit

Evolving the design. The author divided the bungalow into quadrants, with corner cabinets for storage (above left). Next, he sketched in the bathroom and a short hall (center). This left an unattractive freestanding closet in the center of the house. Duffy refined the idea of central storage to a circular closet and a curved hall (right).

partitions for the shelves, parts of which also had to be mated with the central post.

Making these doors was surprisingly easy. The three rails are the only bent sections, and the only differences from straightforward door-building are that you have to bend the rails, feather the flat panels slightly and adjust the layout so you can worry the joints between rails and stiles closed. These operations are shown in the drawings on the next page.

The top and center rails of these doors are 5 in. wide, while the bottom rail is 7 in. wide. I made each rail of three ½-in. thick plies of butternut, each of which was steamed for an hour in a box and then bent over a form that I glued up out of boards cut to the proper curve. I used winding sticks to make sure that the form curved smoothly without any twist (drawing A).

I left the middle ply a couple of inches long on each end so I would have ready-made tenons. I also cut the center ply ½ in. narrower than the outside plies on the top and bottom rails. When the rails are glued in place, its narrow width forms a groove to hold the panels. The center rail needs grooves to accept panels on both sides, so I left the middle ply a full inch narrower than the outer ones. In the center of each middle ply, I also cut a notch 1½ in. deep and big enough to accept a stile. When the rails were glued up, this notch became a mortise for the center stile (drawing B). The butternut was roughsawn, so I cleaned up the field of the panels and roughed out the bevels with a scrub plane, a plane with a convex iron that will remove a lot of stock very fast. Then I jack-planed the field, which left a lovely surface, and finished raising the panels with a panel plane. The doors' stiles were cut slightly thicker and wider than their finished dimensions would be so that I could properly locate their grooves and then plane them to conform to the curve of the rails, and to fit the jamb exactly. I built the jambs in my shop and prehung the doors there.

More curves—There were two other curved elements in the woodwork I did for the house: the concave corner cabinet in the kitchen and the circular bathroom vanity (photos at left). Both of them required tighter bends than the doors, and no amount of feathering would have made flat panels fit. I had to fabricate curved panels as well as curved rails. To do this, I built a cradle out of ¾-in. plywood and 2x2s that established the right curvature (drawing C). Then I laid up 1x3 strips of butternut, after determining the appropriate bevel angle for their edges, as the drawing shows. I cut the bevels on the table saw. The bevel on each strip needed some work with a plane so that it would fit tightly next to its neighbor, stave-fashion. Once the curve looked smooth, I glued the strips together. This work took a lot of time and pa-

More curved built-ins. **The curve became a unifying element in the remodeled bungalow. The kitchen corner cabinet (above left) and the bathroom vanity (left) are variations on the theme. Unlike the doors of the large circular closet, which have curved rails and flat panels, the corner cabinet and vanity have curved panels as well, because of their small radius.**

Making and joining the rails and panels. **The rails of the curved doors are glued up from three ½-in. plies, steamed and bent over a form (drawing A, next page). The middle plies are cut long, leaving ready-made tenons for joining the rails to the stiles. They are also ripped narrower than the outer plies, and are positioned to form grooves for the panels to ride in. The middle ply of the center rail is ripped extra narrow to accommodate both top and bottom panels. Notches in the middle plies create mortises for the center stiles (B).**

Gently curving doors don't require curved panels. Flat panels are simply feathered to fit their grooves. The curved panels of tightly curving doors were assembled on a specially fabricated cradle (C).

tience, and I did a lot of dry-fitting before I got everything to mate. I cleaned up the surfaces of the curved panels with scrub and compass planes, and raised the panels, which were still in the cradle, with the panel plane along the grain, and a small-radius gouge and chisels on the curved ends.

While I was working on this house with its beautiful wood and admittedly unusual design, I felt a strong pull toward what I call quirky work, the eccentric tour de force that all too often has no practical purpose. My temptations led to some reflection on the role of craftsmanship. I began to take a fresh look at my ideas about moldings. I have no formal training in either design or woodworking, but it was clear to me that changes in plan or elevation are not to be taken lightly. I began to wonder about all this graceful static that Chippendale and his friends were so fond of. Was there something to hide? Did the wealthy need a performance every time a horizontal met a vertical? Did they need to avoid recognition of a change? What is this blur in the corners?

I concluded that the best work shows tasteful restraint. Consider the Gamble house by Greene and Greene. For my money, its magnificence is diminished by the extreme amount of thought and action devoted to every feature. No surface was left alone, no plane unaltered. Even rafter ends and door stiles make powerful statements. The house is loved for its honesty, and rightly so, but it is a bullying honesty. I decided on subtle, spare decoration where I wanted any decoration at all. For instance, along the edges of the doors' stiles and rails, I cut a simple bead, which I think makes each panel look less like a picture in a frame, and also results in a gentle play of light and shadow that adds to the warmth of the house.

This job also taught me that you don't need vast spaces to create an interesting design. In fact, I think that a small room or house offers greater possibilities for good design, because there's no space to waste. By introducing angles or curves and by paying attention to details and craftsmanship, you can create a jewel-like richness and a space that feels much larger than it really is. I'm happy that the very center of this little house turned out to be the best place for storage. □

Tom Duffy is an architectural woodworker who lives in Ogdensburg, N.Y.

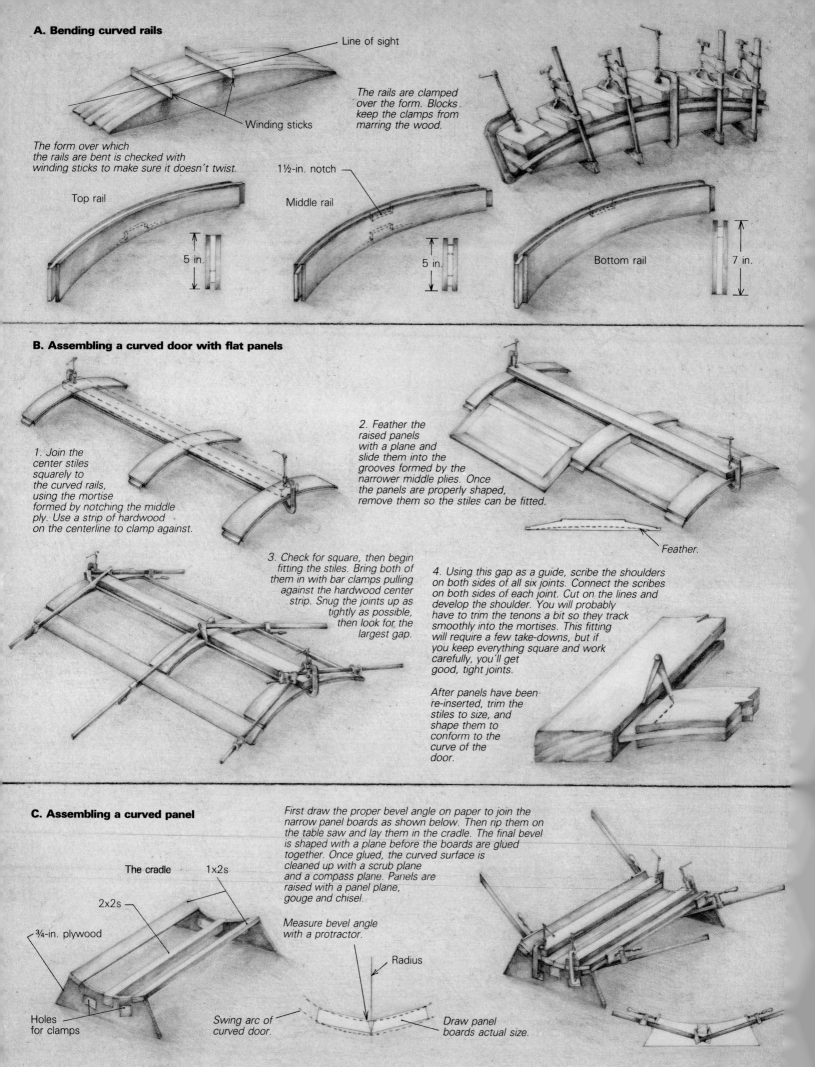

An Elegant Site-Built Door

Build a custom interior or exterior door with common job-site tools and readily available materials

by John Birchard

All it takes are butt joints. Cherry panels floating in a redwood frame warm a room otherwise dominated by a gray carpet and white walls (photo above). The author made the doors on site to match the trim details found throughout the house. The panels in the exterior door (photo below) are narrower, which reduces seasonal swelling and shrinking.

To my eye, nothing adds to the handmade feeling of a home like custom doors do. They are, after all, constantly being touched and used. It stands to reason then that a few extra dollars are well spent on upgraded doors. But a lot of people balk at the idea because custom doors cost plenty. Well, they don't have to.

I make simple, economical doors that also are beautiful. I don't skimp on the materials when I make these doors ($100 worth of materials is about average), and I typically match them with the predominant trim or furniture woods found in the rest of the house. My doors are affordable because they don't take long to make, and they don't require a lot of specialized tools. If you've got a router, a drill, a circular saw and a flat work table, you can build doors right on the job site.

My doors are updated versions of the frame-and-panel style, with simple, clean lines. Instead of using panels that are milled from a single, thick piece of stock, I make the panels for my interior doors out of three layers of plywood (bottom right photo, p. 97). The middle layer is ½-in. birch or mahogany plywood. To this core I glue ¼-in. plywood with the appropriate veneer.

One of the trickiest parts of building doors is joining the stiles and the rails (these are the vertical and horizontal parts, respectively, of the frame). I use two different methods to make this joint. Interior doors don't get the wear and tear of exterior doors, so their frames can be joined with dowels (top left photo, p. 96). This method is less work but not as strong as the loose tenons I use to join exterior doors. More on them later.

On-site door shop—A door-building work station must be outfitted with at least one sturdy work table. The ideal size is around 40 in. by 84 in. I made the one in my shop out of 2x6s. An even simpler version can be constructed easily from ¾-in. A/C plywood (good side up) with a 2x4 frame on edge. Position the table so that you can move around at least three sides of it. On a job site, I prefer to affix one of the short sides to a stud wall and support the other end with 2x6 legs braced with plywood gussets.

This table will be used for laying out the pieces of the door, cutting the joints, planing edges and dry fitting the parts. If you've got room, another nearby table is handy for a tool rest.

One of the most important tools for on-site door building is a heavy-duty ½-in. plunge router. In addition to all the things an ordinary router can do, such as jointing the edges of frame members and cutting slots for panels, a plunge router can also cut loose-tenon mortises for exterior doors.

A half-dozen pipe clamps are essential. I use ¾-in. black-pipe sections that are 48 in. long. This length will clamp the width of most doors, and you can link them with pipe couplings to make clamps long enough to clamp the height of the door if necessary.

Choosing materials—I live on the North Coast of California where redwood is the material of choice for door frames. I use it because of its workability, beauty and resistance to rot and weathering. Folks in other parts of the country will have other local woods to choose from. Pine, poplar (interior only), Douglas fir and oak are all good native species for door making.

I built the interior doors shown in the photo above for a studio addition that has white plastered walls, a light-gray carpet and open-beam ceilings. I think a backdrop like this is a perfect opportunity to showcase the color and figure of varnished or oiled wood. I chose vertical-grain redwood for the door's rails and stiles, and cherry plywood for the panels. Both woods have a compatible reddish tone to them.

I choose the pieces for the stiles carefully. The grain should be straight on both the face and the edges. I avoid stock with twists, large knots or wild grain patterns. Because the rails are much shorter, I can cut up boards with knots and crazy grain to get sound pieces.

If you are working without the benefit of a planer or a jointer, you will have to buy surfaced lumber and use it at whatever thickness it comes. You can use 1½-in. thick lumber for either interior or exterior doors, but 1⅝-in. thick lumber is preferable for exterior doors. Interior doors can be made even thinner—1⅜-in. and 1¼-in. thick doors are common.

Ripping and crosscutting the frame parts—I usually buy 2x6 lumber for doors because the 5½-in. wide pieces are about the right width for the stiles and the top and middle rails. For added strength (and a pleasing look) I use 2x10s or 2x12s for bottom rails. For doors less than 30 in. wide, I'll rip the stiles down to 4¾ in. Most residential locksets and doorknobs are backset 2⅜ in. from the edge of the door, and therefore centered in a 4¾-in. stile. Stiles or rails narrower than about 4½ in. will look out of proportion unless the doors are very narrow.

One of the most important operations in frame-and-panel door building is crosscutting the rails. They abut the stiles, so the crosscuts must be perfectly square, and the rails must all be exactly the same length. Chopsaws (power miter saws) are quite accurate for this work, but many chopsaws don't have a large-enough blade to cut all the way through the wide bottom rail. A carefully tuned radial-arm saw will crosscut door rails, but a table saw with a sliding bed is ideal.

In a pinch, you can also make crosscuts with a fine-toothed plywood blade in a circular saw. To keep the cut perfectly straight and square, use a square for a guide, or use a jig made by screwing two pieces of wood together at right angles. If you're a bit unsteady with the saw, clamp the workpiece and the guide to the bench so that you can use both hands to hold the saw. Check your finished cuts with a square to be sure they're right.

Dowel making. Interior door frames can be held together with dowels. You can make your own dowels with either a quarter-round bit or the half-round bit shown above. The half-round bit should be used in a router table with a fence.

Router jointing. If you don't have a jointer, you can still straighten the edge of a stile if you've got a router. Here, the bearing of a ½-in. by 4-in. flush-cutting bit rides on a plywood straightedge. The bit trims the edge of the stile accordingly.

Jointing with a router—Lumber typically comes from the supplier with rounded corners called eased edges. These edges need to be milled square on the stiles where they abut the rails, and if the edges are a bit warped, they need to be straightened. The jointer is the best tool for this task, but lacking that machine you can accomplish the same thing with a straightedge and a router with a long flush-cutting bit (½-in. by 4-in. spiral fluted bit available from MSM Carbide Tooling & Design, 1232 51st Ave., Oakland, Calif. 94601; 510-532-7669). I make the straightedge out of a strip of ½-in. birch plywood 6 in. wide by 81½ in. long (top right photo, this page). I affix 1x2 blocks to the ends of the straightedge, and a 1x2 fence to one side to cradle the 80-in. long stiles as they are jointed with the router. I drive screws through the blocks and into the ends of the stile to secure it.

As the bit's bearing rides along the plywood straightedge, the cutters mill the edge of the stile. Once I've straightened one edge of the stile, I run it through the table saw to straighten the other edge. If you don't have a table saw, you can add a thin shim to the jig's fence, flip the stile over and joint the remaining edge.

Dowel joinery for interior doors—Dowel joints are quick and easy to make and are plenty strong enough for interior doors. When all the frame pieces for an interior door are prepared, I lay them out on my work table and clamp them together as they will go in the finished door. I center the middle rail (or lock rail) 36 in. from the bottom of the door. The top and bottom rails are flush to the ends of the stiles. I use a square to mark across the joints where the centers of the dowel holes will be. These marks are registration points for a doweling jig (bottom photo, this page).

I put at least three dowels in the bottom rail joints and two each in the middle and top rail joints. I space the dowels evenly and am careful not to put them too close to the inside edges of the rails where the ½-in. deep slots for the panels will be cut later.

Dowel drilling. A doweling jig makes it easy to bore perfectly aligned holes that are perpendicular to the workpiece. Here the jig is being used to drill a rail.

I make my own dowels (top left photo, facing page) out of the same wood as the framework, using a ½-in. half-round bit in a router table. It probably doesn't make any significant difference, but I like to think that making the dowels out of the same wood as the rails and the stiles will mean that the parts will all expand and contract at the same rates. I plane a flat edge onto each dowel for a glue channel.

You can also use 3-in. long birch dowel pins—the kind that are cut to length, beveled and grooved. Don't use smooth dowels because they act like a piston, compressing the air and the glue when you drive the dowels into the holes, and the resulting pressure can split the wood.

Cutting panel slots—After the dowel holes have been bored, I assemble all the parts using a 1-in. long dowel at each joint to align the frame members. I sand the dowels so that they easily slide in and out of the holes. When the parts are all flush to one another, I clamp the frame pieces together while I rout slots in the rails and the stiles for the ½-in. plywood core panels.

I use a ¼-in. slot cutter for this operation (bottom left photo, this page). I set its depth so that it cuts completely to one side of the centerline on the edge of the frame pieces. Then I flip the door over and rout it again. By cutting two ¼-in. slots from both sides of the door, I end up with a perfectly centered ½-in. wide slot for the panels.

Cutting slots this way leaves rounded corners in the bottoms of the slots where the rails meet the stiles. This can be dealt with by rounding the corners of the plywood core panels (top photo, this page) or by using a chisel to chop the corners square. I find it easier to round the corners of the plywood for interior doors. For exterior doors, I think it is better to chop the corners of the slots square to minimize air leakage.

While the framework is still clamped together, I measure carefully from bottom of slot to bottom of slot in both directions to get the sizes for the plywood core panels. I cut these panels from a cheaper grade of plywood, like shop-grade mahogany or birch. I cut them about ⅛ in. small in both directions so that there won't be any danger of the panel holding the framework apart if the frame pieces swell. To make sure everything fits, I dry assemble the door with panels in place.

Glue up—When everything is ready for assembly, I spread the pieces on the workbench in an exploded fashion. I prefer to glue doors with a urea-resin glue like Weldwood. It's a brownish powder that comes in a can and is mixed with water. This is a weather-resistant glue suitable for exterior doors, but I use it for all my doors because it has a slow setup time—about eight hours at 75°. Doors are complicated enough to glue up, and white or yellow aliphatic-resin glues can start to set up while you are still struggling to get all the pieces aligned. Furthermore, urea-resin glue doesn't creep, which is the tendency of pieces to move slightly after the glue has cured. Creep can occur with white and yellow glues because of their relative flexibility.

I brush glue into all the dowel holes and onto all the meeting surfaces of the joints before in-

Radiused and beveled core panels. Cutting the slots in the frame with a router makes for rounded corners. The plywood core panels need to be rounded to match. The panels will fit into their slots easier if their edges are beveled with a hand plane.

Slot cutting. The author uses a ¼-in. slot cutter to mill the ½-in. slot for the panel. With the bottom of the cutter aligned with the center of the workpiece, he makes a pass from one side of the door. A second pass from the other side brings the slot to its required width and ensures that it is perfectly centered.

Applying the outer panels. Panels of ¼-in. plywood with cherry veneer cover the core panels. The outer plywood panels are cut smaller than the frame opening to give a ⅛-in. reveal around the edges. The panels are held fast by glue drizzled on the cores and brads at the corners.

serting the dowels. I dip the dowels into the glue before I insert them in the ends of the rails, then I brush glue along their protruding ends before working them into the holes. This way I can be sure all the meeting surfaces, including the dowels, are covered with glue. I work the rails onto one stile first, slide the panels into place, then insert the dowels into the opposing stile. I use a mallet to tap the pieces together, then I use bar clamps when the pieces are engaged.

As the rail ends come together with the stiles, I put clamps on both sides of the door to keep all the pieces in the same plane. Uneven clamp pressure can cause the stiles to tilt toward the side with the most clamps. I lay a straightedge across the assembled door to be sure it is flat, then I wipe away all excess glue with a damp rag and set the door in a warm place to cure the glue. If your shop is cold, try covering the door with an electric blanket or using heat lamps.

Planing and panels—Once the glue has cured, I go over the entire framework with a sharp hand plane to flatten the joints and remove defects. This could also be done with a belt sander, but the hand plane is better for flattening uneven pieces and leaves less work for the final sanding with 100-grit paper in an orbital sander. It is also important to go over the edges by hand, both inside and outside, to round them slightly. This is most easily done before the veneered plywood panels are applied.

I cut the veneered plywood carefully with a fine-toothed blade to avoid tearing out the grain. The pieces should be ¼-in. smaller in each direction than the distance from stile to stile and rail to rail. This will leave an ⅛-in. shadow line (or reveal) all around the panel, which to my eye looks much nicer than a flat panel. The space between the edge of the framework and the veneered plywood panel is so narrow that the core layers can't be seen.

I sand the edges around the outside face of the veneered plywood panel to round them slightly. This is a subtle touch, but it improves the finished feel of the door. This takes a light hand—too much sanding will eat through the veneer to reveal the core layers.

When affixing the veneered panels, I dribble white glue over the core piece, being careful to lay a line of glue near the edge but not so thick a line that it will squeeze out around the panel (bottom right photo, p. 97). Then I tack the panel in place with a couple of small brads near each edge. Later I'll set and putty them.

Loose tenons. While interior doors can be joined with dowels, the abuse experienced by exterior doors requires the broader gluing surface of a spline, or loose tenon. The spline should be a bit narrower than the mortise to allow the pieces to be adjusted during glue up.

Routing a mortise. A bench vise is the best tool for securing a rail or stile as it is routed. Here, the bit is guided by a fence attached to the router's base. A block clamped to the bench provides a broad surface for the plunge router to ride on.

Loose-tenon joinery for exterior doors— Dowels don't make a strong enough joint for exterior doors, especially in the stiles where most of the contact between the dowel and the stile is end grain—a notoriously weak glue joint. Traditionally, doors have been built with mortise-and-tenon joints, where part of the rail (the tenon) extends into a corresponding hole in the stile (the mortise). This is a difficult and time-consuming joint to cut, but with a plunge router you can make an equally strong joint with ease. It's called a loose-tenon (or spline-tenon) joint. To make it, mortises are routed in the edges of the stiles and in the ends of the rails, and a spline is inserted to hold the pieces together (left photo, this page). The spline has vastly more solid gluing surface than dowels do and is quite strong.

To make the loose-tenon joint with the router, you need a ½-in. by 4-in. plunge bit for your router (either the straight or spiral fluted cutters will work, but the spiral bits cut more smoothly). You also need either a fence attachment (right photo, this page) or a self-centering base to keep the bit centered on the work (see sidebar, facing page).

I start a loose-tenon joint by first marking the positions of the mortises on the workpieces. I lay out the framework of the door on the bench, then I pencil across the joints about 1 in. in from the edges of the joint. So where a 5-in. wide rail meets a stile, the mortises would be 3 in. wide. With a square I extend the marks around the edges of the stiles and the ends of the rails.

To cut the mortises I clamp the workpiece in a vise. Lacking a vise, you could use a bar clamp to

Router mortising jig. *Clamped to a rail or a stile, this simple jig will provide a wide, stable surface for the router's base while cutting a mortise.*

- Rail
- Optional screw-on end stop
- 2x8
- Rabbets for self-centering pins on custom router base

hold the piece on the bench. I like to bore a ½-in. dia. hole at the end of the mortise where I'll be starting my cuts because the router bits don't plunge very easily. I plunge the router in about ½ in. deep and draw it along the mortise until I reach the other end mark, then pull the bit out, go back to my starting point and plunge it in again, repeating the process. It may take four to six passes to cut the mortise to a depth of 2¾ in., which is about as deep as you can go with a 4-in. bit. I find that it's easy to control the length of the mortise by eye.

Cutting the mortises on the ends of the rails is a little more difficult. I often clamp a board next to the rail so that it is flush to the end to provide a larger surface for the router to ride on. A jig can also be made from two lengths of 2x8 that are bolted together at one edge (drawing facing page). The workpiece is held between the two pieces by tightening the bolts and with C-clamps at the top so that it is flush to the top edges of the jig pieces.

The splines should be made from the same wood as the frame pieces, with the grain running parallel to the rails. I leave the splines about ¼-in. narrower than the mortises so that the rails can be adjusted up or down a little on the stiles. With a ¼-in. roundover bit in my router, I radius the edges of my spline stock to match the mortise. I typically use a planer to mill ½-in. thick stock for the splines. Lacking a planer, you can also use ½-in. wide rippings cut on the table saw and ganged together to make up the correct width.

The panels for the exterior door look the same as the ones for the interior door, but they are smaller (top drawing, right), and I make them from solid wood. It would be possible to make solid-wood panels as wide as the ones in the interior door, but I don't recommend it. Wider panels swell and shrink more than narrow ones do, especially when subjected to exterior humidity and temperature swings. As a consequence they are more likely to crack when they shrink or burst the door frame when they swell.

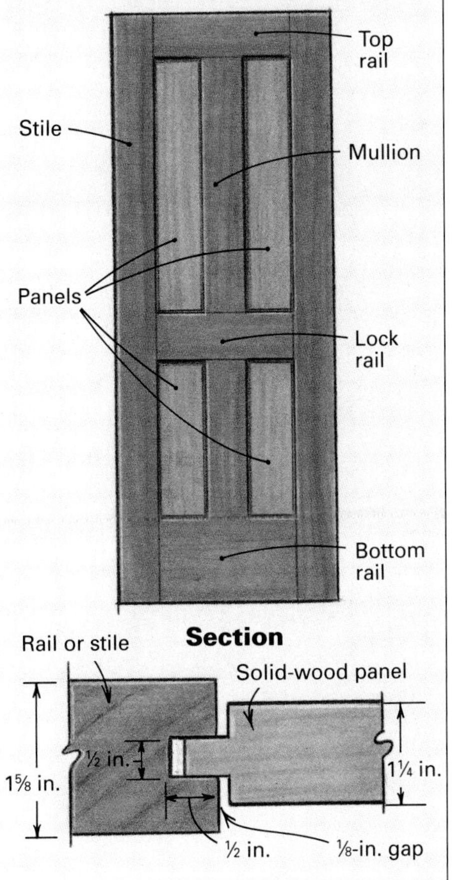

I divide exterior doors into smaller panels by adding a mullion in the center of the frame or by adding additional rails, such as the five-panel door I made for a rustic studio (bottom photo, p. 95). The mullion pieces are joined to the rails with loose tenons, just like the stiles.

I typically make the panels by edge joining pieces of 2x6 and then planing them down to 1¼-in. thickness. Then I use the table saw to cut rabbets around both sides of the panels. The rabbets are ⅜ in. by ⅝ in. so that a ½-in. thick tongue is left around the panel that can fit into the slots in the frame parts (bottom drawing, left).

Good old enamel paint provides the best protection for an exterior door, but if you want to see the grain of the wood on the exterior side, I suggest you use a semiopaque stain and be prepared to do some yearly maintenance. You can renew this kind of finish without a lot of surface preparation. Avoid varnish. It will break down, then you'll have to scrape to get it off. In only a season or two a spiffy-looking varnished door can turn into a disaster. ☐

John Birchard is a woodworker and writer living in Mendocino, Calif. He is the author of Make Your Own Handcrafted Doors and Windows *and* Doormaking Patterns and Ideas *(Sterling Publishers). Photos by Charles Miller.*

Exterior door. *Temperature and humidity fluctuations can cause the panels in exterior doors to swell across their grain. Consequently, it's prudent to add a mullion to the door and make the panels narrower than interior panels.*

Self-centering router base

The self-centering base allows you to position the plunge router quickly in the center of the stock without measuring and using a fence (photo right). To make a self-centering base, I started by duplicating my router base (minus the hole in the middle for the bit) with a piece of ¼-in. thick aluminum plate. Then I chucked a pointed bit in the collet and mounted the aluminum base on the router.

With the bit retracted, I turned on the router and slowly plunged the bit toward the base until the bit just marked the exact center of the base. This gave me the centerpoint I needed to mark a 3-in. dia. circle on the base. I removed the base, drew my circle with a compass and drew another line through the center of the circle across the base. At the two points where the line intersected the circle, I drilled ¼-in. holes using a drill press. They accommodate a pair of posts made of roll pins, which are metal cylinders with a slit along their lengths. They have chamfered ends that allow them to be driven into a hole a little smaller than their diameter. As they are driven, the slits narrow, and the pin is held in place by the spring action. You can get roll pins at a well-stocked hardware store. Once I got the posts installed, I put the custom base back on the drill press and made the ¾-in. hole for the bit.

To use the rig, I lift the router to its retracted position atop its posts and put the base on the workpiece. Then I rotate the router clockwise until the posts snug up against the opposite sides of the work. The bit is now centered and poised for plunging. —*J. B.*

Posts are the secret. **Posts mounted equidistant from the router bit can be used to center a router on the work quickly.**

Installing Mortise Locksets

Whether you drill and chisel or rout with an expensive lock mortiser, cutting a big hole for the case is only part of the job

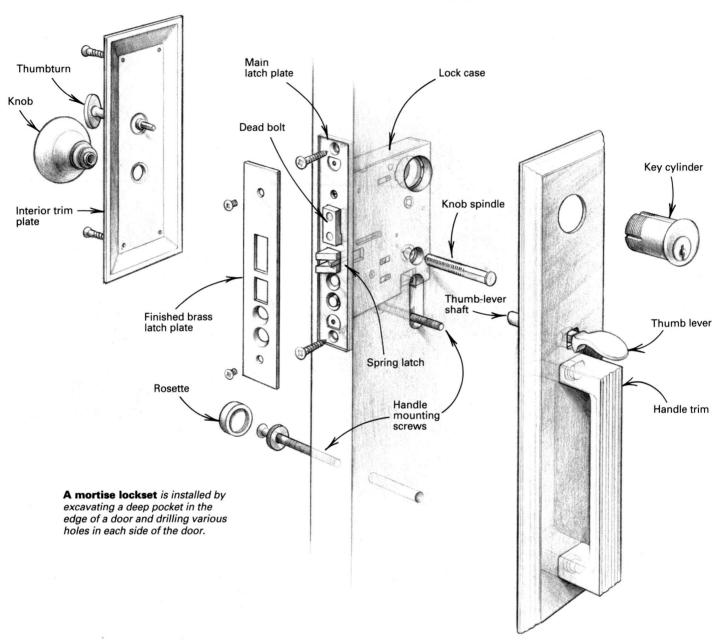

A mortise lockset is installed by excavating a deep pocket in the edge of a door and drilling various holes in each side of the door.

by Gary M. Katz

Mortise locksets (drawing above) are generally considered to be the Cadillacs of door hardware. That's because their larger cases are stronger and wear longer than bored locksets and dead bolts. And because the spring-latch and the dead bolt are housed together, they can be interconnected. For example, some mortise locksets require only one key to retract both latch and dead bolt at once. But mortise locksets are also expensive and difficult to install.

Installing a mortise lockset involves lots of precise drilling. There's the deep pocket, or mortise, in the edge of the door and another mortise in the jamb. There are holes of various sizes on both sides of the door for the knobs, the handles, the levers and the dead bolt, and these holes have to line up precisely. It's possible to do all this drilling freehand, but this article will focus mostly on the tools that speed mortise-lockset installations.

Premortise—The first thing I do when I install any mortise lockset is to figure out the handing of the lock. Holding the lock case with the dead bolt up and looking at the taper of the spring-latch indicates whether the lock was meant for a right-handed door or a left-handed door. Don't unscrew the lock case and try to switch the handing of the lock unless you want to end up with a pile of parts.

Next I measure the backset of the lock case and the width of the door's lock stile. Backset is the distance from the latch plate to the center of the knob-spindle hole. A mortise lockset with a 2¾-in. backset is too wide for the 4-in. lock stile of a standard French door. If the lockset has a 2½-in. backset, I measure the lock stile anyway. Some carpenters install doors by hinging them, then planing the lock stile to fit it. The more a lock stile has been shaved, the easier it is to mortise completely through the wood.

Various handles, knobs and levers are available for mortise locksets. Handle trim sets with thumb levers are common on entry doors, but I always check the trim set that comes with each lockset. You don't want to drill for a handle trim set, then reach into the carton and unwrap a knob. Knob and lever trim sets will not cover the holes drilled for a handle trim set.

Positioning the hardware— I hold the lock case and the handle set against the door to help visualize where the trim will be positioned. If necessary, I adjust the lock height so that the trim is centered on the lock rail. But keep in mind that the most comfortable position for a thumb lever is between 34 in. and 38 in. high. I pencil a mark on the edge of the door at the top and bottom of the case (1).

For a Baldwin mortise lockset, which is the most common brand, I like to make the mortise 6¼ in. high by ¹⁵⁄₁₆ in. wide—slightly larger than the lock case —so that the lock case slips easily into the finished hole. There are several different methods you can use to excavate this mortise.

1. Locating the mortise lock case.

Freehand mortising— In the distant past I cut mortises freehand with a drill and a chisel. Freehand mortising must be done very slowly and carefully. Lock cases are usually ⅞ in. wide, so I bore a series of ¹⁵⁄₁₆-in. holes using a ½-in. drill fitted with a spade bit. Holding the drill firmly with two hands, square to the door (2), I drill the top and the bottom holes first and then make overlapping holes as close together as I can to minimize chiseling later. Some people clamp blocks on both sides of the door to support the thin stock. I always use a piece of tape on the spade bit for a stop guide so that I don't drill too deep. A sharp, wide chisel can be used to shave the sidewall of the mortise (3).

You can also use a lock-boring jig (see p. 106) to make the holes for a mortise. It works like a drill press that you clamp to the door. Using a jig guarantees that you'll drill straight, but it too is a little slow.

2. Drilling a series of overlapping holes.

3. Shaving the mortise clean.

Using a lock mortiser— Several years ago I landed a job that included 12 mortise-lock installations. Because the standard installation fee is between $80 and $100, that job justified the purchase of a Rockwell lock mortiser (now Porter-Cable model 513, Porter-Cable Corp., Youngs Crossing at Hwy. 45, P. O. Box 2468, Jackson, Tenn. 38302-2468; 901-668-8600). A similar tool is available from Bosch (Robert Bosch Power Tool Corp., 100 Bosch Blvd., New Bern, N. C. 28562-4097; 919-636-4200). My mortiser lists for $1,025, but it makes short work of the 40 or so mortise locksets that I install each year.

The tool consists of a router and an adjustable jig that clamps on the edge of a door. A handle cranks the router up and down on the jig, with each pass driving a long carbide-tipped bit deeper into the door. The adjustable depth stop keeps the bit from going through the lock stile, and a dial adjusts the height of the mortise.

Before I start the mortiser, I press a button to engage the gears that drive the bit (4). Then I switch the tool on and crank it up and down the bars on which it is mounted (5). Each vertical pass gradually drives the bit deeper. When the bit reaches the proper depth (about ⅛ in. deeper than the lock being installed), the depth stop automatically disengages the gears that drive the bit, stopping the bit from mortising any deeper. The entire operation, from setup to breakdown, takes about five minutes.

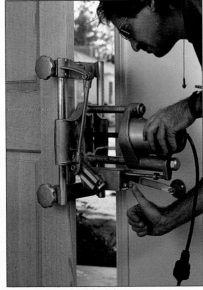
4. Engaging the bit drive.

5. Using the lock mortiser.

Installing the lock case—Sliding the lock case into the mortise, I trace the outline of the latch plate onto the edge of the door, which helps to position my router template (for more on making router templates, see p. 109). I rout a ¼-in. deep mortise for the latch plate (6) and square the corners with a corner chisel.

The lock case won't seat completely in the mortise unless you chisel out a 1-in. high by ¼-in. deep notch on each side of the mortise (7). These notches make room for the latch hinges, which project behind the latch plate.

6. Routing for the latch plate.

7. Notching for the latch hinges.

Trim preparation—The next step is to lay out the locations for the handles, the dead bolt and the decorative plates, collectively called the trim. This means marking each side of the door and then drilling a series of holes through the door and into the mortise.

For accuracy and speed, I made a layout template from a piece of ¼-in. oak (drawing below). I transferred the layout from a manufacturer's paper template, along with the actual lock-spread measurements, and by trial and error (the paper templates aren't all that accurate) I finally made a template that works fine for mortise locksets with 2½-in. backsets, the most popular size that I install.

Securing this template on the face of the door with a spring clamp, I mark with a sharp awl the locations for four of the five holes. I intentionally have not included the fifth hole, for the bottom handle-mounting screw, in my layout template. All exterior trim must be measured. Baldwin manufactures many different handle sets, most of which have different bottom mounting-screw locations. With the trim in hand, I measure from the center of the dead-bolt hole to the center of the bottom mounting screw (8), then transfer this measurement to the door.

I use the same template to mark the interior layout (9). The interior requires only four holes, three marked using the template and one below the template for the bottom mounting screw. Again, I measure for the location of this hole.

When it comes time to drill (10), I start all the through-holes on one side and finish them from the other side. Other holes, like the thumb-turn hole, are drilled only in one side of the door. The template aligns the holes, and drilling from both sides prevents the bits from chipping the back of the door.

On the exterior of the door, I connect the third and fourth holes by carefully chiseling out the wood between them, forming a keyhole mortise (11).

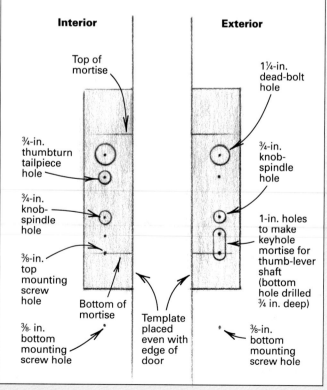

8. Measuring the trim.

9. Marking the layout.

10. Drilling the interior layout.

11. Chiseling keyhole mortise.

Installing the lock—Because the edge of a door is beveled, the lockset's main latch plate (there's also a finished brass plate that fits over the main plate and is installed later) must be adjusted to match the bevel of the door. Otherwise, the case won't sit squarely in the mortise, and the dead-bolt key cylinder won't seat properly against the exterior trim. Screws on the top and the bottom of the lock case allow you to adjust the latch-plate bevel.

Once the lock case fits, you can insert the threaded spindle that will hold the interior knob (12). This spindle threads in from the door's exterior and cannot be installed once the exterior trim is in place.

On Baldwin locksets you have to cut the thumb-lever shaft to match the thickness of the trim and the door you're working on. Hold the exterior trim with the thumb-lever shaft against the linkage plate (13). The shaft is so long that the trim can be more than an inch from the face of the door. Measure that distance, then cut that amount off the shaft with a hacksaw.

With the thumb-lever shaft trimmed and the handle set in place, carefully thread the dead-bolt key cylinder into the lock case (14). Use a key to turn the cylinder all the way in. Both Baldwin and Schlage happen to use the same keyways, so I use a Schlage key to tighten cylinders. I've found that Schlage keys are stronger than Baldwin keys, which I've had snap off inside the lock. On the face of the main latch plate there's a set screw that secures the dead-bolt key cylinder.

Two wood screws mount the lock case in the door; I drill for these screws first. While tightening the screws, I check every function of the lock; one overly tight screw can really foul things up. Now the finished brass plate can be installed over the main latch plate.

Next come the two handle-mounting screws. These screw into the back of the handle from the interior side of the door. The upper screw is hidden behind the interior trim, and the lower screw (15) has a threaded base and a finishing washer that accepts a rosette, which covers the screw. Again, while tightening these and all other screws, I check all the functions of the lock, making sure the dead bolt throws and retracts with the key, that the thumb lever retracts the latch, and that the latch springs back.

To install the interior trim, I start by engaging the tailpiece of the dead-bolt thumbturn into the lock case. Next, I thread the handle-set knob or lever against the trim plate (16) tight enough to hold the trim plate in the right position. After drilling pilot holes and installing the brass screws to secure the trim, I check again to see that the lock functions smoothly. Then I tighten the interior handle until it causes the latch to hang up in the case. I back off one turn and tighten both set screws.

12. Threading in the knob spindle from exterior.

13. Measuring the thumb-lever shaft.

14. Use a key to tighten the dead-bolt cylinder.

15. A mounting screw secures bottom of handle.

16. Threading the knob onto the spindle.

Installing the strike plate—On a Baldwin mortise lockset, the strike plate is mortised into the jamb 1⅛ in. below the top of the lock's latch plate, with the large hole for the dead bolt up (drawing right). I use a router template to make the strike-plate mortise (17), setting the router's depth for the thickness of the strike plate, the dust bucket and the security insert—a steel plate installed behind the strike plate that strengthens the strike plate.

To check the location of the strike plate, I measure the edge of the door from the exterior face to the interior side of the latch. This is the distance that the interior side of the strike plate's latch opening should be from the doorstop. If the strike plate is too close to the doorstop, the latch will not enter the latch hole; if the strike plate is installed too far from the stop, the lock will latch, but the door will rattle in the jamb.

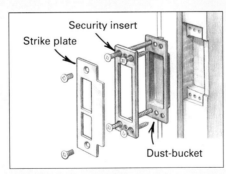

17. Making a shallow strike-plate mortise.

After routing the shallow mortise for the strike plate, I drill pilot holes for the mounting screws. Then I can position my 1-in. wide by 3⅝-in. long mortise for the dust bucket. I use a 1-in. Forstner bit to drill overlapping holes for this mortise, which I make freehand. It's easier to overlap holes with a Forstner bit than with a spade bit. Then I clean up the sides of the deeper mortise with a chisel.

With the dust bucket, the security insert and the strike plate screwed into place, I pull my cords out of the opening, sweep aside the sawdust and slowly swing the door shut. I hold my breath as the door hits the stop. I wait for the latch to fall into the strike. Clink-klunk. The dead bolt next. Click-klunk. Then my lungs relax, my blood flows, and I'm ready for another door. □

Gary M. Katz is a carpenter/contractor and writer in Encino, Calif. Photos by Rich Ziegner.

Installing Locksets

You can do the job freehand with a chisel and a drill, but jigs and routers are faster

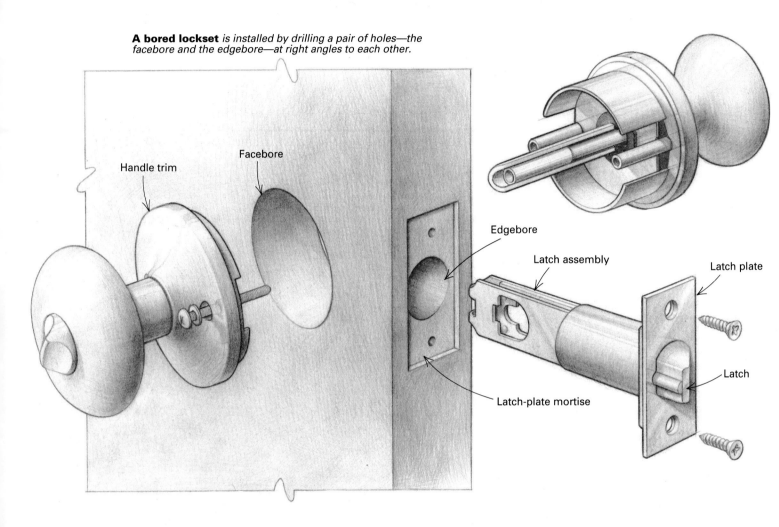

A bored lockset is installed by drilling a pair of holes—the facebore and the edgebore—at right angles to each other.

by Gary M. Katz

I don't know about you, but in my Top 10 Causes of Dizziness and Nausea on the Job, "Sheathing a Roof in a Stiff Wind" and even "Working Beneath a Hungover Stonemason" don't rank as high as "Drilling the Edgebore Freehand in an Expensive Door." There's just not much room for error. As little as 3/16 in. stands between you and a large monetary loss.

I've been installing door hardware for years, and in this article I'll explain the process, from drilling the door to tightening the last mounting screw. Along the way I'll show you how to install a lockset if all you've got is a drill, a hole saw, a spade bit and a chisel. But I'll also talk about the specialized tools I use that help me work faster and that spell relief from dizziness and nausea.

There are two basic kinds of locksets: bored locksets and mortise locksets. Bored locksets (drawing above), the more common of the two, get their name from the fact that you install them by boring holes in the door. They come in a variety of price ranges and are relatively easy to install. Mortise locksets are big metal cases that contain both the door latch and the dead bolt. They are expensive and tough to install because they have to be mortised into the edge of the door. I explain that process on pp. 100-103. But here I'll concentrate on bored locksets.

Gary M. Katz is a carpenter/contractor and writer in Encino, Calif. Photos by Rich Ziegner.

Locating the bores—Whether you're installing a lockset, which is spring-loaded, or a dead bolt, which operates only with a key or a thumbturn, the same procedure applies. First you locate and drill two holes—one big hole through the face of the door for the handle, called the facebore, and one smaller hole in the edge of the door for the latch assembly, called the edgebore.

Most lockset manufacturers supply a paper template to align the two holes you'll have to drill. The template also helps you find the backset—the distance from the leading edge of the door to the center of the facebore. To be safe, I measure the lockset, regardless of the template. Residential locks usually have a 2⅜-in. backset, and commercial locks normally have a 2¾-in. backset. A lock with a deep backset cannot be installed on a single-panel door or a French door with a 4-in. lock stile because the handle trim will overhang the back of the lock stile. Even worse, on a French door you risk drilling into the glass.

Another important dimension is the lock's distance from the floor. For a bored lockset, draw a line square across the edge of the door somewhere between 35 in. and 36 in. If you've got a raised-panel door with a rail near lockset height, called a lock rail, center the lockset with the lock rail.

Fold the template to fit on the edge of the door (1). Some templates have a perforated area that you pop out so that you can see the line you drew on the door's edge. Place the template on this line and mark the center of the facebore on both sides of the door. Then find the center of the edgebore by dividing the line on the door's edge in half (drawing left).

1. Mark the facebore with an awl.

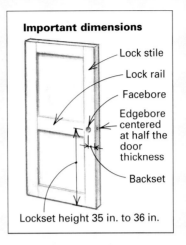

Important dimensions
- Lock stile
- Lock rail
- Facebore
- Edgebore centered at half the door thickness
- Backset
- Lockset height 35 in. to 36 in.

Freehand drilling—You'll need a 2⅛-in. hole saw to drill the facebore. Hole saws vary in quality and in price, but don't sacrifice quality for a few bucks. A continuous-cylinder hole saw with a mandrel bit is superior to the hole saw with an adjustable arm extending from a centerpoint. And don't be tempted by those inexpensive multisize sets with the concentric blades that nestle inside each other; they're awkward, time-consuming and don't even drill a round hole.

When you make the facebore (2), drill until the point of the hole saw's mandrel bit emerges from the other side of the door (3) and then finish from the other side (4). Drilling from both sides means less worries about your drill splintering the door.

Now switch to a 1-in. spade bit for the edgebore (5). I drill the edgebore second because, with the facebore drilled, sawdust won't build up and trap the bit in the door. Put your foot against the door to keep it still, hold the drill level, sight the drill bit so that you are drilling straight into the door and let 'er rip. Don't worry, you're drilling only about 1¼ in. into the door, and you probably won't allow the bit to wander out the face (gulp!).

4. Finish facebore from other side; use hole to guide mandrel bit.

2. Start facebore, aligning hole saw with center of lock rail.

3. Stop drilling when tip emerges from opposite side.

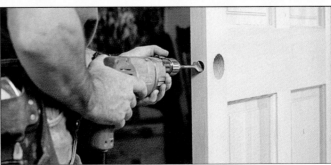
5. Hold drill level and square to drill edgebore with spade bit.

Doors 105

A boring jig—Hole saws are slow, and spade bits tend to wander, causing dizziness and nausea, especially when you're drilling an expensive door. The remedy? A lock-boring jig (6). This tool comes in a kit that has everything you need to install a lock—drill bits, markers, templates and the jig itself. The jig works a lot like a drill press that you clamp to a door. It assures accurate backsets, exact facebore and edgebore alignment and straight, worry-free edgeboring. The boring jig I use is from Classic Engineering (4344 Artesia Ave., Fullerton, Calif. 92633; 714-521-4087), but Templaco Tool Company (295 Trade St., San Marcos, Calif. 92069; 619-471-2550) also makes a good one. Although the Classic Engineering kit is less expensive than Templaco's, you can get either model for around $300.

Instead of a hole saw, the boring jig comes with a 2⅛-in. spur bit for drilling facebores. A spur bit has a short, triangular tip at its center and small carbide cutters around its circumference, like a sophisticated hole saw. Spur bits cut quickly and accurately, and they leave a clean hole. An arm on the front of the boring jig holds a bushing through which you slip the bit's shaft. Opposite of this arm is a clamp pad with a 2⅛-in. hole in it. When it clamps on a door, the pressure of the jig's clamp pad prevents the bit from chipping around the back of the 2⅛-in. facebore. A reducer ring threads into the 2⅛-in. hole and prevents chipping when boring 1½-in. facebores for Schlage dead bolts. The jig also has spring-loaded backset stops that are easily adjusted for a 2¾-in. backset or a 2⅜-in. backset.

6. The boring jig works like a drill press that you clamp to a door.

Using the jig—A 1-in. edgeboring bit is always in my boring jig. I position the jig on the door by touching the point of the 1-in. edgeboring bit to the pencil mark I drew for the edgebore. The jig itself has a small wing nut that adjusts for the thickness of the door and centers the edgebore. I slide the jig forward so that the backset stops are resting firmly against the edge of the door. Then I clamp the jig snug on the door by tightening a big wing nut at the front of the jig.

The strange thing about a boring jig is that you have to chuck up the bits after the jig is clamped on the door. My kit came with a quick-release driver to make it easier to set up the bits in the drill. But the quick release could not be disengaged while the drill was spinning, and I like to move quickly. To solve this problem, I fit my ½-in. drill with a standard ⅜-in. socket extension to drive the bits. I can pull my ⅜-in. extension off the bit while the drill motor is still spinning, which I couldn't do safely with the quick-release driver supplied with my boring kit. I didn't have to make any modifications to the drill or the bits; the extension fit into the ½-in. chuck, and the bits in the boring kit have boltlike hexagonal ends.

I drill the facebore first (7), and then I pull the bit straight through the door (8). Removing the spur bit from the jig saves the tip of my edgeboring bit from colliding with the spur bit while I'm drilling the edgebore (9).

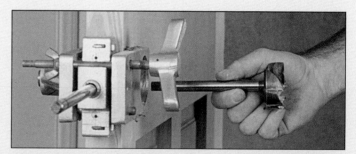

8. Remove the spur bit from the jig before drilling the edgebore.

7. Drive the faceboring bit with a drill fitted with a socket adapter.

9. A boring jig ensures straight, worry-free edgeboring.

Chiseled mortises— For the latch assembly to sit flush in the edge of the door, you'll need to mortise for the latch plate. Then you locate and drill a hole in the jamb to catch the latch, and mortise the jamb for the strike plate. Put the hardware together, and you're done.

With the facebore and the edgebore drilled, slide the latch assembly into the edgebore, square it up and trace around the latch plate with a sharp pencil (10). Use a scratch awl and a knife to score the outline of the latch-plate mortise. On the top and the bottom, use the knife to make a clean cut. On the sides use the scratch awl (11); it won't get caught up in the grain and stray off the mark.

With a ¾-in. chisel, score the wood in about ¼-in. increments down the length of the mortise. The scores, which are made with a single hammer blow to the chisel held at 45°, go square across the grain. Scoring helps prevent wood tearout as you cut the mortise. Tap the chisel down from the center to the bottom of the mortise (12 and 13), then finish it by tapping up from the center to the top of the mortise. A latch-plate mortise should be about ³⁄₁₆ in. deep. Remember, always keep the chisel's bevel against the wood. Insert the latch assembly and check the fit; shave a little more if necessary.

Once the latch plate is flush with the edge of the door, close the door so that the latch is touching the edge of the jamb. Find the center of the latch (you can eyeball this one) and mark it on the side of the jamb. Open the door and square this mark across the face of the jamb. Center the hole in the strike plate on the line and position the strike plate so that it's centered between the door stop and the edge of the jamb. Trace around the outside of the strike plate as you did the latch plate, but then also trace around the inside of the latch opening (14). Before chiseling, drill a 1-in. dia. hole ⅝-in. deep into the jamb, right in the middle of the tracing you made of the latch opening (15). Then use the same chiseling technique to mortise for the strike plate. Drill pilot holes (16) for all the mounting screws so that you won't split the jamb or the door.

10. Insert the latch and trace the latch-plate outline.

11. Use an awl to score the two edges with the grain.

12, 13. Chisel out the scored mortise from center to bottom and then finish it by chiseling up from the center.

14. Trace the strike plate and the latch opening.

15. Drill the latch hole in the middle of the penciled outline.

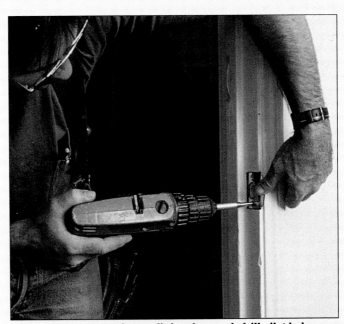
16. To prevent screws from splitting the wood, drill pilot holes.

17. A template tacked to the door has an opening that guides a straight router bit set to cut a shallow latch-plate mortise.

18. Set the corner chisel in the mortise and strike the retractable plunger blade to square off the rounded corners left by the router.

Routed mortises

Just as my days of drilling doors freehand are behind me, so too are my days of chiseling mortises for latch plates and strike plates. Instead, I use a router and some homemade templates (see sidebar facing page). My templates require the same router setup as most professional hinge-mortising templates—a ½-in. straight-cutter bit (Porter-Cable #43614) and a ⅝-in. O. D. router collet (a removable template guide in which the bit spins). Because the collet is ⅛ in. wider than the cut diameter of the bit, the template openings are ⅛ in. larger than the door hardware.

I attach the appropriate latch-mortising template to the door (17) and center it by eye. However, locators are available that fit into the template opening and into the bores to position the templates. I place my router against the template so that the router bit is inside the template opening, then I turn on the motor. This practice helps save the template from nicks. I keep the router collet in contact with the edges of the template opening, then I hog out the center. When I am finished, I try to remove the router without nicking the sides of the template. I don't always succeed. I use a corner chisel (Porter-Cable #42234) to square up the rounded corners left by the router (18) and plastic auto-body filler to fill nicks in my router templates.

My boring kit came with a device called a center marker (a steel cylinder with a sharp point centered on its face) that's used to find the exact position for the strike hole on the jamb. I slide the center marker into the edgebore (19), point out, and close the door against the doorstop. I stick my finger into the facebore and push the point of the locator into the jamb (20). This leaves a small dimple, which I carefully circle with a pencil. The center marker assures that the strike hole and the latch will align properly; the pencil mark guarantees that I won't drill a 1-in. strike hole in a gravel dimple or a nail hole. Imagine shutting a finished door and seeing the strike 3 in. above the lockset. Right. Dizziness and nausea.

I chuck up a 1-in. Forstner bit to drill the strike hole. A Forstner bit, with its razor-edge cutting rim and its interior pair of chisellike lifters, leaves a clean, flat hole. For a lockset I try not to drill completely through the jamb unless I'm installing a dust bucket—the molded plastic or brass insert that provides a finished background within the strike hole. Actually, dust buckets do more than just look good; they prevent loose material behind the jamb from slipping into the strike hole and fouling the lock operation. With the strike hole drilled, I attach a strike-plate template to the jamb and follow the same procedures I used when mortising the door.

The rest is a simple matter of screwing together the hardware. I drill pilot holes and install the latch plate (21) and the strike plate first, then I link up the handle trim. Remember not to set the screws too tight, which could interfere with the operation of the hardware. I check every operation of a lockset at least twice—once against dizziness, once against nausea. □

19. Insert the center marker, point out.

20. Push the marker to dent the jamb.

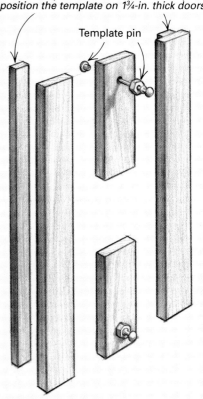

21. Install the latch before the knobs.

Making router templates

You can buy manufactured templates for routing latch- and strike-plate mortises, but the templates I've seen are plastic, and if you nick one, you might as well throw it into the recycling bin. I made my templates from hardwood (photos below). If I nick one, I fill the nick with auto-body filler, and it's as good as new.

Instead of cutting a hole in a larger piece, I get a crisp template opening by gluing up strips of hardwood (drawing below). Because the collet on my router is ⅛ in. wider than my ½-in. straight-cutter bit, the template openings are ⅛ in. larger than the latch and strike plates. For example, I made the template for a 1-in. by 2¼-in. latch plate by gluing two 1⅛-in. wide hardwood strips between another pair of hardwood strips about 12 in. long. I spread the 1⅛-in. strips 2⅜ in. apart—the length of the latch plate plus the thickness of the collet.

After trimming the ends of the template, I drilled a hole at the top and the bottom and installed template pins (Bosch #83018). A template pin is a retractable brad set in a sleeve that's threaded on the back. It allows you to tack the template in place and then remove it easily. You can buy template pins individually or by the dozen at just about any serious tool outlet.

Because I work mostly on 1¾-in. thick doors, I screwed two narrow hardwood strips 1¾ in. apart on the back of each latch-plate template, one strip on each side of the opening. This modification allows me to position the template quickly with only one pin.

On my strike-plate templates, I use a ⅛-in. steel strip on the doorstop side to complete the opening. I notched some of these templates with a jigsaw to create the shape of the flange on a T-shaped strike. I grooved these same templates lengthwise and put another steel strip in the groove. This removable strip covers the notch and lets me use the same template to mortise for a T-strike or a rectangular strike.

Latch plates are generally a little thicker than strike plates. Therefore, I made my strike templates ¹⁄₁₆-in. thicker than my latch templates so that I don't have to change the depth of my router. Actually, all my templates are based on the depth of hinge mortises. I often mortise for latch plates, strike plates and hinges at the same time, so I've linked the thickness of my strike and latch templates to the thickness of my hinge templates. With this system I rarely alter the depth of the ½-in. router bit. —G. K.

Strike-plate templates.

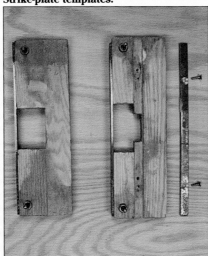

Latch-plate templates.

Latch-plate template. *Hardwood strips are glued so that the opening matches the size of the latch plate plus ⅛ in. for the router collet.*

Narrow strips fastened to the back help position the template on 1¾-in. thick doors.

Template pin

Homemade Hardware
Turning scrap metal into hinges, hooks and grilles

by Chip Rosenblum

I always resented working hard on a piece of custom woodwork only to come up empty when looking for well-designed hardware at a reasonable price. So I learned to make my own. Making my own hardware is expensive in time, but it's affordable (shopping the local recycling center is very cost-effective) and emotionally gratifying. Although I've made various pieces of hardware from a variety of materials, hinges of brass and copper are what I make most often (photos right).

The first step in making any piece of hardware is understanding what you want it to do and what you want it to look like. A hinge, for example, is anything that can be attached to one fixed and one movable surface. A piece of leather nailed across two boards is an adequate hinge, but might not meet your aesthetic or functional demands. Metal hinges in their simplest form consist of one loop set over a pin with a way to attach each to a different surface; these are called hook-and-eye hinges. The butt hinges we're used to seeing on doors are a stack of meshing cylinders, attached to a pair of leaves, with a pin through them to allow rotation.

For my hinges, I use copper plumbing pipe for the pins and no-hub copper repair couplers for the knuckles (left photo, facing page). The repair couplers will slip-fit over the pipe to allow rotation and are rigid enough to support a door (if it's not too heavy). They come in standard 1-in. lengths, or can be bought in 12-in. lengths and cut to whatever size your design requires. I use a minimum of three knuckles for stability. The outside ones are attached to one leaf and the center one to the other.

The design process for creating hardware is a balance between artistic freedom and the functional and aesthetic requirements of the work it will serve. You need to look without preconceptions at objects and see other functions in their form. Parts from an old toilet-seat hinge can be made into unusual hooks. A polished storm grate from an old gymnasium makes a wonderful heating grille. Brass covers to old water meters, cut-up sections of handrail brackets, gears, bushings and even parts of old musical instruments are some of the things that I've used. Recycling centers will usually let you "shop" the bins if you explain what you're doing.

Cutting and drilling—While I have a small woodshop, my metalworking experience was

Among the parts used to fabricate these homemade hinges are, from top to bottom, a gear cut in half, a footrail bracket from a fancy bar, and the cover to an old water meter.

limited to cutting and soldering plumbing parts. But I reasoned that if I could hacksaw brass and copper fittings, I could cut up and rearrange pieces for hardware. I found that hand tools are adequate for this work, but they're tiring to use and tough to control (they leave wavy cut lines). I decided to try using my bandsaw.

The specification sheet for the ¼-in. bimetal blade I started with said it was designed to cut nonferrous metals (brass, copper, aluminum). I clamped a brass gear to a board, set the rip fence to bisect the gear, and with great trepidation slowly fed the gear through the blade; it cut like butter. Now I use the same blade for cutting both wood and metal (middle photo, facing page).

Hand drills or portable power drills are sufficient to make and countersink holes. But as with the bandsaw, a drill press offers safe control of depth and hole location. To make screw holes, I use a woodworking countersink (right photo, facing page), which is fine for nonferrous metals, and then follow that with a standard twist bit $\frac{1}{32}$ in. larger than the shank diameter of the screw I'm using.

Brazing brass and copper—Before I could begin brazing, I needed a safe surface that could handle the high degree of heat needed for working with metals. A piece of backer board sitting on a bench was a good beginning. I covered the backer board with full-size firebrick from my local building-supply store (large photo, p. 112). I also stocked up on half-thickness firebrick to stabilize and brace the pieces of metal while I worked with them.

My next problem was finding the proper heat source to do the job. In *Metal Techniques for Craftsmen* (Oppi Untracht, Doubleday and Co., Inc., 1968), I looked up the melting points of pure copper (1981.4° F), brass (an alloy of copper and zinc, 1650° F) and bronze (an alloy of copper and tin, 1550° F to 1900° F). The theoretical flame temperature of my propane plumbing torch is about 3500° F, so I figured I was in business. I stood with a propane torch for about half an hour watching a pile of copper pennies turn red, shrivel a little, but not melt. The intensity of the flame was insufficient to keep up with the conductivity of the metal.

At the hardware store I found a MAPP (Methylacetylene propadiene stabilized) air torch, si-

Rosenblum made one pair of hinges from brass footrail brackets, no-hub repair couplers and ½-in. copper pipe.

Using a scrap of 2x4 to support the bracket, he cut it into four pieces on a bandsaw fitted with a ¼-in. bimetal blade.

Rosenblum clamped the pieces securely and used a woodworking countersink in a drill press to countersink each screw location.

milar to my propane torch, but marketed with the word "brazing" on the box. Good, I thought, I can get this done now. No such luck. The MAPP-gas torch might be sufficient to braze thin-gauge steel, but could not melt or braze any reasonable mass of metal.

I realized then that I would have to invest in an oxyacetylene (or oxymapp) welding outfit. The flame temperature of acetylene in oxygen is 5600° F and mapp/oxygen is 5200° F. With either setup, brazing, welding and cutting are possible. A small, portable welding outfit with tanks, regulator, tip and carrier cost me about $320. Different size tips are available for working with everything from fine jewelry to large welds.

It is critical to wear welding goggles with filter lenses (for you and for any observers—your supplier will recommend the correct shade for the work you are doing). Brazing rod is available in different diameters—thicker rod is used for heavier work. I find that $^3/_{32}$-in. flux-coated rod ($4.45/lb.) is a good general purpose size.

Brazing, sometimes called braze-welding, is similar in principle to soldering electrical or plumbing parts—you're using a nonferrous alloy to join two pieces of metal. The term *welding* refers to joining pieces of steel with steel welding rod. In both techniques the parts themselves must be hot enough to melt the rod and prevent a cold (unfused) joint. The instructions with the welding outfit will tell you how to set up the tanks and adjust the pressure of the regulators.

Never use any oil or flammable solvent to lubricate the valves of gas equipment, as this, combined with oxygen under pressure, can cause an explosion. The adjustment of the welding flame is the critical part of assuring the right temperature and environment for successful brazing or welding. Just as a candle or Bunsen burner will have an inner and outer cone of flame, so does a welding torch. The adjustment of the inner cone of the welding flame is controlled by the proportion of fuel gas to oxygen.

To get started, open the fuel-gas knob at the tip (the regulators have previously been set), and ignite it. You'll see a yellow, smoky flame, resembling an undernourished candle. By slowly turning on the oxygen supply you'll see a shorter, bluer flame (remember, you're wearing goggles now). At this point you'll have three indistinct cones of flame, called a carburizing (still too much gas and carbon) flame. As the oxygen is turned up, a defined, inner cone of deeper blue will emerge. Continue turning up the oxygen pressure until there is no "halo" of a third cone inside the outer cone and in front of the inner cone of flame. You now have what is known as a neutral flame. This is what you want for welding. But when used for brazing with brass filler rod, a neutral flame has a tendency to "burn out" the zinc, leaving pits. Increasing the oxygen just a little to a slightly oxidizing flame reduces this tendency and encourages smoother flow of the filler rod.

A major consideration while brazing or welding is adequate ventilation. These processes produce noxious and/or toxic fumes, which must be exhausted to the outside. It is not enough to filter the fumes under an unvented hood that simply recirculates the same air into your environment. They must be ducted outdoors or you must work in an open space. I set up just inside my garage with the garage door open and a fan behind me to force the fumes outdoors.

Getting metal to the right temperature before applying the brazing rod is a matter of trial and error. Because "found" brass objects are of unknown composition and may melt at the same temperature as the brass brazing rod, a gentle yet persistent touch is needed. My first attempts were underheated, resulting in a good-looking joint that promptly fell apart as it cooled because I had managed to melt only the flux without the brazing rod ever fusing to the parts.

With increasing confidence and more heat, I then proceeded to turn my carefully aligned parts into what looked like residue from a meteorite re-entry. Somewhere between these two extremes is the key, and with time and practice I can now braze brass with a reasonable expectation of success.

Grinding and polishing—Assembling the parts, I discovered, is only the beginning. The hardware is almost unrecognizable when it emerges from the brazing process. It's black, flux-encrusted, scaled and lumpy. It looks horrible (top right photo, next page). The processes of finishing and polishing seems endless, but a logical sequence of steps makes it easier. The first thing I do is drop the piece in water for a day or two, as this seems to help any remaining flux to "bloom." This means that the flux becomes white and fluffy, instead of glasslike, and the residue is easier to remove.

Then I use a grinding wheel (bottom right photo, next page) and a hand-grinder with stones and abrasive-impregnated rubber points to remove the bulk of the overcontoured brazing rod (when brazing, you build up the joints more than you need to allow material for finishing and polishing). At this point, if I no-

Goggles with filtered lenses protect Rosenblum's eyes from the light of the torch, and the open garage door behind him, combined with a fan, prevent dangerous fumes from building up as he brazes the hinge leaves to the knuckles (couplers).

Once the brazing process has been completed, the assembled hinge, though functional, is far from beautiful and far from finished.

The brazed joints are shaped and smoothed on a grinding wheel, then polished on rag wheels, dressed with polishing compound.

tice any voids in the joints, I rebraze those sections and start the grinding and polishing processes all over again.

The next step is to polish the metal on a rag wheel dressed with a selection of polishing compounds (this is also a great device for sharpening knives and chisels). A motor-driven polishing wheel must be used with care. The 4-in. scar on my left thumb is testimony to how a wheel can grab if you feed material too aggressively. You should set up the polishing wheel so that as you face it, the top of the wheel rotates toward you. If you imagine a horizontal line through the middle of the wheel parallel with the floor, the work should be held below this line. This way if the wheel does grab, the work will be thrown down and away from you. If you hold the work above this horizontal line, the wheel will immediately catch the work and throw it back at you.

The polishing compounds applied to the wheel do the work, so a gentle pressure against the wheel is more effective and safer than pushing aggressively into the wheel. Like sanding wood, polishing metal is a process of removing coarse scratches with finer scratches. A progression of grit sizes will produce a nicely finished piece of work (photos, p. 110).

A different polishing wheel should be used and marked for each grit, as even a single piece of coarser grit can ruin the next stage with scratches that are too deep. For the same reason, the bench that you're working on should be swept and the work washed between grits. It may be tempting to skip a step, but cutting corners will actually create more work, just as a jump to too fine a grit will not adequately "polish out" the previous scratches.

I wear a dust-mask respirator while using a grit sequence of 80, 150, 220, 320, 400, and even finer grit-polishing compounds, such as tripoli and finally green rouge to finish and polish. These materials and the polishing points and wheels are available through jewelry and lapidary houses, or a local plating and polishing shop may sell you enough to get started if you explain what you're trying to accomplish.

Practice—I get a lot of pleasure from making something that I can't buy. And searching for raw materials also justifies many stops at garage sales, flea markets and scrapyards. If you can work with wood, you can work with metal—all it takes is practice. Every time I create a new piece of hardware from someone else's trash, it's a little better than the last one I made, and I remind myself of the old joke: "How do I get to Carnegie Hall?" And the answer is "Practice, man, practice." □

Chip Rosenblum lives in Columbus, Ohio, and is known as "the brass man" by the folks at the local recycling center.

Door Hardware
Getting a handle on locksets, latches and dead bolts

by Kevin Ireton

Walter Schlage, a German immigrant working in San Francisco, obtained his first patent in 1909. The idea came to him late one night, as he unlocked his front door and reached in to turn on the lights. The patent was for an electrified door lock that automatically turned on the lights inside a house when the front door was unlocked.

You can't relate the history of the lockset industry without talking about Schlage, Dexter and Yale, men whose companies are still in business. But today, more than 50 other companies now manufacture locksets and dead bolts. From 18th-century reproductions to ergonomically designed lever handles, more door hardware is available today than ever before.

Most door hardware performs at least two basic functions—latching and locking—and many aspire to a third function—looking good. The term lockset refers collectively to the complete latch bolt assembly, trim and handles (knobs or levers). A latch bolt is a spring-loaded mechanism that holds a door closed and may or may not have a lock incorporated into it. A dead bolt, on the other hand, is not spring-loaded and can only be operated with a key or a thumbturn.

Rim locksets and mortise locksets—Originally just wooden bolts that slid across the opening between door and frame, rim locksets were the first broad category of locksets to evolve. They developed into the surface-mounted wrought iron and brass cases common in colonial America.

Eventually, locksmiths realized that doors would look better if the locksets were out of sight. So they mortised the rim lockset into the edge of a door to create the mortise lockset, which prevailed throughout the 19th century. Mortise locksets are still widely used, especially in commercial installations. In residential construction, they're used for exterior doors more often than for interior, though models for both uses are available.

Some people consider mortise locksets to be the best type of door hardware available (photo above). Manufacturers can build more

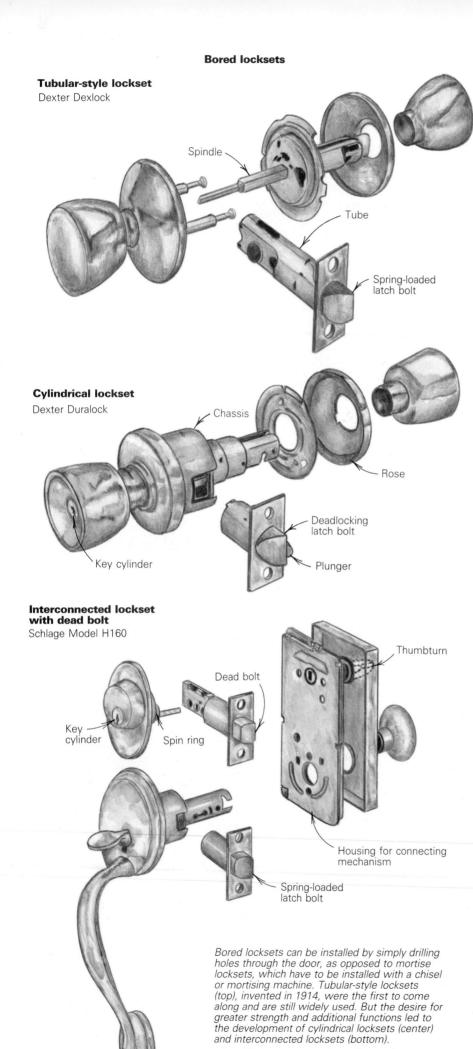

Bored locksets

Bored locksets can be installed by simply drilling holes through the door, as opposed to mortise locksets, which have to be installed with a chisel or mortising machine. Tubular-style locksets (top), invented in 1914, were the first to come along and are still widely used. But the desire for greater strength and additional functions led to the development of cylindrical locksets (center) and interconnected locksets (bottom).

strength and longer wear into the larger cases than they can into the bored locksets that I'll discuss later. Also, because the latch and dead bolt are housed together, they can be interconnected to offer a variety of functions. Turning the inside knob on a mortise lockset, for instance, retracts both the latch and the dead bolt—no fumbling for the dead-bolt thumbturn if you're in a hurry to get out. The push buttons, called stop-works buttons, are located below the latch on most mortise locksets and determine whether the latch bolt can be retracted by the outside knob. This type of lockset is also easily adapted for use in extra-thick doors. All that's needed is a longer spindle for the knob.

Aside from the fact that they cost more than other types of door hardware, mortise locksets carry the disadvantage of being tough to install. Cutting a mortise 6-in. long, 4-in. deep and nearly 1-in. wide in a door that's only 1¾-in. thick is tricky and time-consuming. The standard method involves drilling a series of closely spaced holes in the edge of the door and chiseling to clean out between them.

To ease the job, Porter-Cable Corporation (P. O. Box 2468, Jackson, Tenn. 38302-2468) makes a tool called a lock mortiser (model 513). It clamps to the edge of the door and has an integral motor that holds a long mortising bit, much like a router does. Turning a crank on the side of the unit moves the bit up and down along the door and advances the depth automatically. Once it's set up, the lock mortiser works fast, but at a suggested list price of $975, you'll have to mortise a lot of doors to pay it off.

Bored locksets—In 1914, a Michigan hardware manufacturer named Lucien Dexter realized that installing mortise locksets on every door in a house amounted to overkill—nobody needed that much hardware just to keep the pantry door shut, so Dexter invented the tubular-style lockset.

The tubular-style lockset is basically a spring-loaded latch bolt, housed in a tube, that mounts in the edge of a door (top drawing, left). It's operated by a spindle that runs through the end of the tube. A smaller, simpler mechanism than the mortise lockset, the tubular-style lockset is also much easier to install. All you have to do is drill two intersecting holes in the door, one through the face—the crossbore—and the other through the edge—the edgebore (for more on installing bored locks, see p. 76).

The drawback to the tubular-style lockset is that most of the working parts are contained in the narrow tube, so there isn't room to incorporate many functions or a very secure lock. Walter Schlage overcame this problem with his next invention in about 1924, when he developed the cylindrical lockset. Like the tubular-style lockset, the cylindrical lockset incorporates a spring-loaded latch bolt mounted in the edge of the door. But instead of just a spindle passing through the tube, a bigger crossbore hole is drilled (usually 2⅛-in. dia.), and a large chassis is inserted through the door. The cylindrical lockset and the tubular-style lockset are classified together as bored locksets because both are installed by boring holes.

Before the invention of the cylindrical lockset, latching and locking were performed by separate mechanisms. But the large chassis on his new invention gave Schlage enough room to install the key cylinder right in the knob, so that locking and latching could be combined in one mechanism (for more on key cylinders, see sidebar this page). Schlage also developed the button-lock for the inside doorknob, an idea he got from push-button light switches in use at the time.

Among later refinements to the cylindrical lockset was the deadlocking latch bolt. The familiar trick of retracting a latch bolt by slipping a credit card between the door and jamb is known in the trade as "loiding a lock," *loid* being short for celluloid. The deadlocking latch bolt thwarts this practice. With the door closed, a spring-loaded plunger, located alongside the latch bolt, is held in by the strike plate, and the latch bolt can't be retracted. This requires a reasonably conscientious installation. Otherwise, the plunger can slip into the hole for the latch bolt, which defeats the purpose.

Dead bolts—"Don't rely on a lockset for security." I've heard that from a number of people in the industry. Projecting out from the door like it does, a key cylinder in a doorknob is vulnerable to being smashed or sheared off. If you want security, you have to have a dead bolt. Some insurance companies even offer you a 2% or 3% discount on homeowner's insurance if you have dead bolts on all your exterior doors. Still, I was reminded by a locksmith that, "One-hundred percent security is not available on this planet. Anything that's manmade can be defeated. By using dead bolts, you're simply trying to make access more difficult."

Dead bolts should have a full 1-in. throw, which means the bolt will extend 1 in. into the door jamb. At one time, shorter throws were common, and some companies may still make them. But if the dead-bolt throw is less than 1 in., it's easy for an intruder to prize the door jamb away from the door and free the bolt.

Many companies run steel inserts through the center of their dead bolts so that if someone tries to hacksaw them, the inserts will roll under the blade instead of allowing it purchase to cut. It's a great idea in theory, but no one I asked had ever heard of a burglar hacksawing a dead bolt; there are too many easier ways to break into a house. Peace of mind, I learned, is a major factor in selling home security.

Most dead bolts include a key cylinder outside and a thumbturn inside the door. But dead bolts are also available with a double cylinder, which means you need a key to unlock it from either side of the door. Double-cylinder dead bolts are used in doors that have windows so an intruder can't break the window, reach in and unlock the door. In emergencies, though, they can pose a safety hazard, and their use is restricted by code in some areas.

The weak spot in a dead-bolt installation is the strike—the metal plate mounted in the door jamb that the dead bolt goes into. It should be mounted with 2-in. or longer screws that penetrate the trimmer stud behind the jamb. Some companies supply long screws with their dead bolts, but some don't, and often you have to angle the screws to sink them solidly into the framing. But now a few manufacturers have begun to offer dead-bolt strike plates with holes that are offset toward the center of the door jamb.

Many dead bolts come with a metal or plastic housing, called a strike box or dust box, that installs behind the strike and keeps the dead bolt contained inside the jamb. Installing it requires extra work with a chisel, and I'll admit to throwing away more strike boxes than I've installed. I thought their only purpose was to make the hole in the jamb look neat. But it turns out that strike boxes serve a couple of important purposes. For one thing, they keep debris from getting into the strike hole and jamming the bolt. But

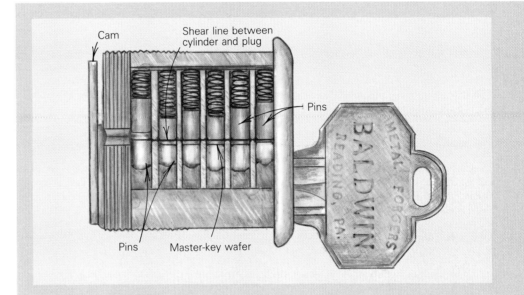

Pin-tumbler cylinders

The oldest known lock, found at the ruins of the palace of Khorsabad in present-day Iraq, is over 4,000 years old and operates on the same principle as most door locks today—the pin-tumbler. This lock and others like it were operated by a wooden key shaped something like a large toothbrush, but with wooden pins in one end instead of bristles. When the key was inserted and lifted up, the pins on the key aligned a set of pins inside the lock that allowed a wooden bolt to be retracted. This type of lock was common in Egypt and is called an Egyptian lock. Subsequent civilizations developed new locks based on different principles, and pin-tumblers weren't used again until 1848 when Linus Yale "reinvented" them.

Yale developed and patented a simple key-operated lock based on the pin-tumbler principle. His son later refined the lock by encasing it in a small metal cylinder with a rotating internal plug. Although plenty of locks have been invented since—combination locks, push-button locks, credit-card locks and even electronic locks—the pin-tumbler cylinder is still the heart of most door locks used today.

How it works—A series of holes, usually five or six for a door lock, are drilled through the top of the cylinder into the plug. At least two pins of different lengths, plus one spring, are inserted into each hole. Some cylinders have brass pins, but the best cylinders use nickel silver, which is a harder metal than brass and assures that the key will wear out before the pins. By varying the combinations of pin lengths, a manufacturer can key any given cylinder literally millions of different ways.

The line between the plug and the cylinder is called the shear line. Any misalignment of the pins across the shear line prevents the plug from rotating in the cylinder, and hence, the lock from unlocking. But when the right key is inserted in the keyway, the pins slip into the gullets of the key's serrated edge, which raise them to just the right height, aligning their ends along the shear line (drawing above). The plug is then free to turn in the cylinder, and when it does, a cam attached to the back of the plug throws the bolt.

A master key is one that will work two or more cylinders that are keyed differently. Cylinders accommodate a master key by means of master-key wafers inserted with the regular pins. These wafers create additional points where joints between the pins align at the shear line and the plug is free to move.

Most key cylinders are assembled by women because the pins are very tiny, and women, with their smaller hands, can assemble cylinders much faster than men can. *—K. I.*

their main function is to serve as a depth gauge. If the dead bolt is less than fully thrown, only ¾ in. for instance, it will easily spring back into the door when you tap it—not a very secure situation. Installing the strike box ensures that the dead bolt can be fully thrown.

The large beveled housing surrounding the outside of a dead-bolt cylinder is called a spin ring. It protects the key cylinder from being removed with a pipe wrench or locking pliers. When buying a dead bolt, be sure to get one that has a solid metal insert for the spin ring. Hollow rings can be crushed easily.

Weiser Lock (5555 McFadden Ave., Huntington Beach, Calif. 92649), makes a combination lockset and dead bolt called the Weiserbolt. It works like a standard lockset, until you turn the key in the lock or twist the thumbturn. Then the latch bolt extends a full 1 in. into the jamb. But while the Weiserbolt is more secure than a standard lockset, it isn't as secure as a separate dead bolt because the key cylinder in the projecting knob is still vulnerable. It does, however, provide additional security without requiring an extra hole in your door.

Interconnected locksets—In order to combine one of the functions of a mortise lockset with the easy installation of a bored lockset, the Schlage Lock Co. introduced the interconnected lockset in 1967. It's an entrance lockset that links the latch and dead bolt so that both will retract by simply turning the inside handle. The mechanism connecting the dead bolt and latch bolt is contained in a thin rectangular housing that mounts on the inside face of the door (bottom drawing, p. 114). Interconnected locksets cost more than separate locksets with dead bolts, but they are more convenient to use.

Interior locksets—Locksets for interior doors don't usually have a keyed lock, but may instead have a simple button lock. These are called privacy locksets and are used for bedrooms and bathrooms. Privacy locksets have an emergency release so that a locked door can be opened from the outside either with a special tool that comes with the lockset or with a small screwdriver.

A passage lockset is one with no locking function at all; turning the knobs or lever handles on either side of the door will always retract the latch. Some manufacturers offer a special closet lockset that has a knob or lever handle on the outside, but a simple thumbturn on the inside. These are usually available only in more expensive lines, because if you're paying $30 or $40 for a solid brass doorknob, you may not want waste it inside a closet. A dummy knob is a handle with no lockset and is often used on the stationary half of a pair of French doors.

Quality and standards—Many companies produce a broad range of locksets that vary, not just in style and finish, but in quality. Two locksets made by the same company may look exactly alike behind the cellophane windows on their packages, but one might cost $25, and the other, $70.

Among the factors that determine the cost of a lockset are whether the key cylinder is made of turned brass (which can be machined to close tolerances and is long-wearing), a zinc die casting, or plastic. The steel parts inside a good grade of lockset are treated for corrosion resistance; cheaper ones aren't. One locksmith I talked to suggested picking up the locksets to gauge their weight. The heavier one, she told me, is probably better. Someone else suggested working the mechanism and trusting my instincts about which feels smoother.

It's likely that the $25 lockset is a tubular-style and the $70 lockset is a cylindrical. Following World War II, manufacturers developed the ability to install key cylinders in tubular-style locksets. But in general, tubular-style locksets don't meet the same standards that cylindrical locksets meet.

The Builder's Hardware Manufacturer's Association, Inc. (BHMA, 60 East 42nd St., New York, N. Y. 10165) sponsors a set of lockset standards published by the American National Standards Institute (ANSI) and certifies locksets according to those standards. Products from companies that subscribe to the standards (not everyone does) are selected at random and without notice and are tested for strength, performance and finish by an independent laboratory. The grading that results is rather involved, but a few examples will give you some idea of what goes on.

Bored locksets (ANSI A156.2) are certified in three grades. Grade 1 means that the lockset is suitable for heavy-duty commercial use. Grade 2 means it's suitable for light-duty commercial use. And grade 3 means it's suitable for residential use.

In one test, a lockset is attached to a door that's opened and closed by machine at a rate of 10 times per minute. A grade 1 lockset

The photo above shows the stages a lever handle goes through in the hot-forging process. Once cut to length, pieces of bar stock, called billets, are heated to 1,400° F. The red-hot billets are placed in a die, slammed by a forging press, and the shape of the lever emerges. Excess brass (flashing) squeezes out around the edges of the forging and is trimmed on a punch press. The lever then goes from grinders to polishers to buffers and finally to the lacquer spray booth.

has to be operational after 600,000 cycles, grade 2 after 400,000 and grade 3 after 200,000. In another test, a 100-lb. weight is swung into a closed door from a given height to see if the latch bolt will bend or break, allowing the door to open.

Should you consider installing grade 2 locksets on a house? Not necessarily. Grade 2 hardware will stand more abuse than grade 3, but the difference might only be whether it takes five minutes to break into the house or three minutes. In either case, the weak link is most likely to be the door and the jamb, not the hardware. Grade 2 locksets will certainly be more durable than grade 3. But whether they will see enough use in a house for the difference to become apparent is another question.

Not all manufacturers participate in the certification program, and not every product of the participating manufacturers meets the minimum standards. To make matters worse, most companies don't advertise their certification on packaging. They do, however, include it in their catalogs, so if you're interested, get a catalog. Or better yet, contact the BHMA and for $2.50, get a copy of their "Directory of Certified Locks and Latches."

Reaching for the brass knob—Brass, an alloy composed of 60% to 70% copper mixed with lead and zinc, is the premier material for door hardware. It offers the right combination of strength, workability and corrosion resistance. According to builders and hardware dealers I talked to, some of the finest brass knobs and handles generally available are made by the Baldwin Hardware Corporation (841 Wyomissing Blvd., Reading, Penn. 19603). If you're converting the room above your garage into a rental unit to help meet your mortgage payments, Baldwin locksets and handles probably aren't for you. Their solid brass hardware is expensive, but considered worth the cost by those who can afford it. I visited Baldwin's plant to find out why.

The heart of Baldwin's operation is hot forging, the same process used by blacksmiths. Raw material, purchased from brass foundries in the form of bar stock, is cut to length, and the pieces, or billets, are conveyed into furnaces and heated to 1,400° F. The red-hot billets are then placed in dies and whomped into shape by a huge forging press. Excess material called flashing seeps out around the forging and is later trimmed on a punch press (photo, facing page).

Baldwin touts hot forging over the sand casting used by other companies by pointing out that the resulting pieces are denser, stronger and smoother, lacking pit holes and air pockets. But in the context of residential use, they aren't necessarily more durable or more secure. In this price range, the difference comes down to aesthetics—how it looks and feels, which is why Baldwin takes such pains with polishing, buffing and finishing.

The biggest challenge faced by Baldwin and by any manufacturer of expensive hardware lies in the nature of custom homebuilding. Door hardware is one of the last things to be installed in a house, and if the project is over budget, hardware is a common target for cost-cutting. To learn what a cost-cutting homeowner or builder might use instead of solid brass, I visited the Dexter Lock Company (300 Webster Road, Auburn, Ala. 36830). Dexter does sell a designer series of forged brass door hardware, but I wanted to learn how they make the ubiquitous hollow doorknob that most of us reach for every day.

At Dexter, doorknobs are formed on a sixteen-stage transfer press. The brass is 8 in. wide and .028 in. thick and is fed to the press from coils as big as wagon wheels. After cutting it into 8-in. squares, the machine transfers the brass from one die to the next, forming the familiar shape a little more each time and trimming off the excess. Toward the end of the process, the knob is filled with fluid and expanded by hydraulic pressure to create the final shape. The whole process takes less than a minute.

The big difference between solid brass and hollow brass doorknobs ends at that point. Both get much the same white-glove treatment as they're passed from polishers to buffers and then conveyed on racks through the lacquer spray booth and drying ovens.

Exotica and where to find it—The Schlage Lock Co. (2401 Bayshore Blvd., San Francisco, Calif. 94134) recently introduced something called a Key N' Keyless Lock. It's a lockset with optional dead bolt that can be opened without a key simply by twisting the knob left and right in sequence. A British company, Modric, Inc. (P.O. Box 146468, Chicago, Ill. 60614) makes lever handles in 355 colors, with cabinet hardware and bath accessories to match (photo below, center).

Valli & Colombo (P.O. Box 245, 1540 Highland Ave., Duarte, Calif. 91010), an Italian company, recently introduced a designer line of door handles developed for the disabled (photo below, left). Meroni, another Italian company, makes a push-button doorknob (distributed by Iseo Locks Inc., 2121 W. 60 St., Hialeah, Fla. 33016) that opens when you squeeze it (photo below, right).

Normbau Inc. (P. O. Box 979, 1040 Westgate Dr., Addison, Il. 60101) and Hewi, Inc. (7 Pearl Ct., Allendale, N. J. 07401) both make colorful nylon-coated door hardware that's tough and won't corrode or tarnish.

You probably won't find these products in local hardware stores or lumber yards. Look in the Yellow Pages under "architectural or builder's hardware," or go to a locksmith. If you are interested in antique or reproduction hardware, check out the listings in the *Old-House Journal Catalog* (Old-House Journal Corporation, 69A Seventh Ave., Brooklyn, N. Y. 11217. $15.95, softcover). You can also write to the Door and Hardware Institute (7711 Old Springhouse Road, McLean, Va. 22102-3474) for a copy of their *Buyer's Guide, 5th Edition*. It will cost you $35, but you'll get the most comprehensive list of door-hardware manufacturers that I've seen. □

Kevin Ireton is an associate editor with Fine Homebuilding.

Designer handles for the disabled (left), made by Vali & Colombo, ergonomic levers in 355 colors from Modric (center) and push-button doorknobs from Meroni (right) are just a few of the European door handles recently introduced in the United States.

FINISHING TOUCHES

More Than a Door

Southwestern folkhouse entry. Michael Sandrin designed and built this Sante Fe home using simple architectural elements indigenous to New Mexico. Photo by Michael Sandrin.

Bull's-eye door lights and transom. The door on this stone cottage in Kirkcudbright, Scotland, sports a bowed window with bull's-eye glass. Photo by Alasdair G. B. Wallace.

Traditional New England entryway. Typical of old New England seacoast architecture, this entry can be found on Nantucket Island. Photo by Charles Wardell.

Brass and glass in Boise. The glass and metalwork for this entry were done by Noel Weber of The Classic Sign Studio in Boise, Idaho. Photo by Charles Miller.

Unique screened doorway. Screens are used to keep insects out while letting cool breezes in. However, this unusual screen door designed and built by architect Daryl E. Hanson also softens the entrance to Hanson's Minneapolis home. Photo by Kevin Ireton.

FINISHING TOUCHES

A Welcome Home

Here's a gallery of doors surrounded by various materials, including stone, stucco, and ferrocement (look closely at the door in the lower right corner). Clockwise from left: Miami, Fla.; Fairfield, Conn.; Miami, Fla.; San Diego County, Calif.; Cape Cod, Mass.; Miami,

FINISHING TOUCHES

Photo: Andrew Gulliford

Fla.; London, England; Aspremont, France. All photos by Nancy Hill except where noted.

Doors 121

FINISHING TOUCHES

Sculptor Tom Williams got into architectural work after a builder asked him to carve some kitchen cabinet doors, and his business has expanded since to include other doors as well. Exterior passage doors are glued up from solid 1¾ in.-thick mahogany. Mahogany blocks provide stock for raised carving. Where a view is called for, tempered glass let in to the back of the door is used to meet local building codes. Standard locksets fit well; Williams just has to leave a flat area for mounting.

FINISHING TOUCHES

Carved Doors

Readers are invited to send photographs and a description of traditional or contemporary details to Finishing Touches, Fine Homebuilding, Box 355, Newtown, Conn. 06470. We pay for items we publish.

Garage doors present design problems of a different order. The raised portions are carved from ⅝ in. mahogany, and the flats are ¼ in. mahogany veneer. All the carving is done first in the shop, then taken to the site and applied to a standard flat-panel garage door with construction adhesive and finishing nails. —*Mark Feirer*

FINISHING TOUCHES

FINISHING TOUCHES

Corner Copia

Finish carpentry is the hallmark of homebuilding craftsmanship, and one good place to find it is around doors. The Thomas Ruggles house (photo left), in Columbia Falls, Maine, was built in 1818. Completely restored, the house is now open to the public every day from June 1 until October 15. Douglas fir trains speed over the doorways at Charles Miller's house (above). The curve of the "wheels" was made by running the wood on edge through the tablesaw and stopping it short. Jatoba, a Brazilian wood, provides the raw material for Craig Savage's casing (below). The returns were mitered, doweled and glued to the head casing. Plugs cover screws that secure the casing to the wall. Granite isn't a traditional casing material, but David Wilson likes it anyway (photo right, above). The granite was lag bolted to the framing through slots cut with a circular saw and a diamond blade. The Douglas fir at Tom Faxon's house (photo right, below) is finished with 3 parts turpentine, 2 parts linseed oil and 1 part satin polyurethane. The blend penetrates deeply and gives the casing a durable finish that's easy to repair.

INDEX

A
American National Standards Institute (ANSI): lockset standards of, 116

B
Bandsaws: metals on, 110, 112
Bathrooms: sinks for, circular pedestal, 93
Birchard, John: *Doormaking Patterns and Ideas*, cited, 99
Bits:
    hole-saw, for locksets, 105
    router, for jointing, 96
Blomberg, Nancy J.: *Navajo Textiles*, cited, 86
Brass: hardware from, making, 110-12
Brazing: process of, 111
Brody, J. J.: *Mimbres Painted Pottery*, cited, 86
Butternut *(Juglans cinerea):* for cabinets, 92, 93

C
Carving:
    layout for, 84-86
    painting, preparation for, 86
    See also Latches: sculpted.
Casings:
    corner-block, 78, 79
    and drywall, considerations for, 77
    edges of, easing, 79
    for entryways, jamb for, 26
    installing, with prehung doors, 68-69
    jambs of, 70-73, 77
    mating of, to walls, 79
    mitered, 78, 79
    nail holes in, filling, 79
    nailing, 78
    publication on, 77
    relieving, 77
    removing, 79
    square-cut, 78, 79
    staining, 79
    stock for, 77
    tools for, 77
Caulk: adhesive water-proof, source for, 32
Chamberlain, Samuel: *New England Doorways,* cited, 20
Chisels: corner, retractable, 108
Circular saws:
    laminate shoe for, 55-56
    worm-drive flush-cut, source for, 52
Closets: circular, 92
Color Putty Co.: source for, 79
Columns:
    fluting, 15
    See also Pilasters.
Copper:
    hardware from, making, 110-12
    as roofing, 26, 27
Cornices: over entrances, 24, 25

D
Design: overcare in, 93
Disabled: door handles for, 117
Door and Hardware Institute: *Buyer's Guide 5th Ed.:* cited, 117
Doors:
    associations for, 13
    batten, making, 40-41
    beadboard for, 38
    bevels for, 56, 58
    books on, 47, 80
    bucks for, plywood wedged, 58
    cabinet, curved, 93, 94
    classical interior, building, 80-83
    clearances for, 55, 56
    commercial,
        choosing, 8-13
        sources for, 13
    crosscutting, methods for, 55-56
    custom, source for, 86
    exterior,
        designing, 99
        finish for, 99
        glue for, 97
    fasteners for, 40, 41
    fiberglass, 8, 12-13
        finishing, 12-13
        repairing, 13
    finish for, 10-11, 12, 26, 38, 59
    fitting, 27, 58
    framing for, 57-58
    French,
        hardware for, 30-31
        with rabbeted edges, 31
        retrofitting, 29-33
        sliding, 30
    gallery of, 118-19
    gang of, installing, 76
    gauge blocks for, 77-78
    gluing up, 38, 39
    hand of, 58, 70-71
    hanging, 33, 70-77
    with plumb bob, 62-63
        scribing anchor for, 55
        worm-drive flush-cut saw for, 52
    hardware for, 51
    insulated, making, 38-39
    latches for, hand-forged, 85
    lockset jig for, 54
    making, 25-26, 38-39, 95-99
    molding for, 26
    ornate, gallery of, 120-21
    paint color for, 10
    paneled, 26
        circular, making, 93-94
        nonsplitting, 9
    pediment for, with urn, 83
    pocket,
        circular, 92
        constructing, 89, 90
        early, repairing, 91
        frame kits for, 87-90
        latches for, 90
        manufacturers of, 89
        pulls for, 90
    prehung,
        installing, 60-63, 64-69
        ordering, 60
    prices of, compared, 12
    production hanging of, 52-56
    pulls for, hand-forged, 85
    reveals for, 77
    with rounded glass panels, 82-83
    screen, 48-51, 119
    sculpted, 122-23
    sidelights for, 21, 22, 23
    with Southwestern motifs, 84-86
    steel, 8-9, 11
        dents in, repairing, 12
        edges of, 12
        finishing, 12
    storm, making, 51
    thresholds for, *in situ* replacement of, 42-44
    trimming, 58
    veneered, over laminate, 8-9
    wood, insulated, 9, 10
        R-value of, 10
Douglas fir *(Pseudotsuga menziesii):* for trim, 125
Dowel joints:
    for doors, 96-98
    on drill press, 81
Dowels:
    drilling for, 96
    making, for doors, 96
Drill presses:
    doweling jig for, 81
    metal drilling with, 110, 111
Drilling: with jig, 96

E
Ellis, George: *Modern Practical Joinery*, cited, 23
Entablatures: constructing, 17-18
Entryways:
    arched chamfered stone, 49
    book on, 20
    casing for, 27
    cornices for, 26, 27
    designing, 14, 24, 25
    diagram of, 16
    doors for, 24, 25-27
        double-, history of, 14, 16
    entablature for, 17-18
    formal, details of, 24
    gallery of, 118-19
    Greek Revival, building, 20-23
    moldings for, 15
    pedestals for, 15
    pediments for, 18-19
    pilasters for, 15, 26-27
    wood for, 25

F
Fillers:
    recommended, 79
    spackling compound for, 79
Finishes:
    penetrating-oil,
        coloring, 86
        maintenance of, 86
    varnish, polyurethane, marine two-part, 43
    wax and oil, applying, 86
    See also Paint.
Flutes (grooves): making, 23
Frame-and-panel: for doors, 80-83, 97, 98
Framing:
    for French-door retrofit, 29, 31, 33
    headers for, built-up retrofitted, 33
    of jambs, production, 52-56
Friezes: doorway, 16, 17
Furniture: custom, source for, 86

G
Gas lines: and door retrofits, 30
Glues:
    epoxy, for exterior use, 48-50
    exterior, 26
    for exterior doors, 97
    hot-melt,
        for mockups, 37
        types of, 37
Greek Revival: entryway in, 20-23

H
Hardware:
    door, 51
        installing, 59
    for French doors, 30-31
    hand-forged, source for, 86
    making, 110-12
    pocket-door, suppliers of, 88, 90
    See also Hinges. Latches. Locksets.
Hinges:
    for batten doors, 40
    cup, installing, 73-75
    gains for, router bits for, 54, 55

installing, 56, 58-59
maker of, 38
making, 110-12
painting, 86
for prehung doors, 66-67
router template for, 32
springing, 56
templates for, 54, 55
for thick doors, 38
Hodgson, Fred: *Modern Carpentry I,* cited, 23
Holes:
  filler for, colored, 79
  spackle for, 79

I

Insulation:
  MEPS, 11
  polyurethane,
    and ozone depletion, 11
    R-value diminishment of, 11

J

Jatoba *(Hymenaea courbaril):* trim of, 125
Joinery:
  mortises, for hinges, chiseling, 59
  tenons, table-saw jig for, 47
  See also Dowel joints. Mortises. Tenons.
Joint compound: recommended, 79
Jointing: with router, 96

K

Kitchen cabinets: with Southwestern motifs, 85

L

Latches:
  hand-forged, 85
  sculpted handmade, 126-27
Light fixtures: custom, source for, 86
Locksets:
  Baldwin Lexington pattern, 26
  bored, installing, 104-109
  boring jigs for, 106
  brass, 117
  cylindrical, 114-15, 116
  deadbolt, 114-16
    keyless, 117
  for disabled, 117
  hand of, 70-71, 100
  installing, 38, 59, 101-103
    jig for, 54
  interior, 116
  kinds of, 113
  lever-handled, 117
  locating, 117
  mortise, 113-14
  mortiser for, 114
    power, 101
  pin-tumbler cylinder, 115
  for prehung doors, 67-68
  quality of, 59, 116-17
  tubular, 114-15, 116
  Weiserbolt, 116
  See also Latches.

M

Mail chutes: installing, 28
Metals:
  bandsawing, 110, 111
  drilling, 110, 111
  grinding, 111-12
  polishing, 112
  working, book on, 110
See also Brass. Copper.
Mimbres motifs: book on, 86
Miters: clamping, with glued blocks, 37
Moldings:
  books on, cited, 23
  eave crown to raked crown, mitering, 20, 21-23
  for entrances, 24, 26
  glossary of, 16
  making, 23
  with multiplane, 15
  of shaper, 82-83
Mortises:
  for door latches, chiseling, 107
  jig for, 98
  for locksets, routing, 108
  router templates for, 108

N

Nails:
  needlepoint, 27
  removing, tool for, 42
National Wood Window and Door Association: address for, 13
Navajo motifs: book on, 86

O

Old-House Journal Catalog: cited, 117

P

Paint:
  colors for, blending, 86
  endgrain bleeding of, preventing, 86
  primer for, oil-based stain killer, 32
Pilasters: bases of, repairing, 43, 44
Planes:
  compass, 18, 19
  eighteenth-century, reproduction of, 19
  multi-, 15
Plaster:
  cutting, with reciprocating saw, 32
  patching, around new door, 33
Plumb bobs: door hanging with, 62-63
Plumbing: and door retrofits, 30

R

Railings: bracket for, handcarved wooden, 85
Routers:
  bases for, self-centering, 99
  jointing with, 96
  mortises with,
    hinge, 73-75
    jig for, 98
    latch, 108
    lock, 101
  templates for,
    hinge, 32
    mortise, 109
  two-flute hinge-gain bits for, 54
  for weatherstripping groove, 32
  See also Bits.
Rowland, 47
R-value: handbook for, 38

S

Safety:
  in brazing, 111, 112
  in metal polishing, 112
Screening: coated fiberglass, source for, 51
Screens: door, making, 45-48
Shapers:
  cope-and-stick cutters for, using, 82
  moldings on, curved, 83
Shells: over entrances, 24, 27-28
Siding: cutting, with reciprocating saw, 32
Southwestern arts: book on, 86
Stanley Works, The: multiplane 55, using, 15
Stone masonry: for trim, 125
Stools: portable work, 77
Stucco: cutting, with carborundum blade, 31, 33

T

Table saws: tenons on, jig for, 47
Taylor, Lonn, and Dessa Bokides: *New Mexican Furniture,* cited, 86
Templates:
  for door hanging, 73-75
  for hinges, 32
  for lockset mortises, 109
Tenons: loose, for doors, 96-99
Torches: for brazing, 110-11
Transoms: diagram of, 24, 26
Trim:
  bullnose casing for, 61, 62
  casing, making, 80, 82-83
  corner-block, mitered, 36
  custom, from stock components, 34-37
  Douglas fir, 125
  finish for, 125
  granite, 125
  jatoba, 125
  ornate, 124
  removing, tools for, 42
  train, 125
  See also Casings.

V

Victorian: doors, screen, 45-47

W

Weatherstripping:
  for doors, 58, 59
    thresholds of, 59
  router-groover for, 32
  spring-type, installing, 58, 59
Western Wood Moulding and Millwork Producers Association: casings booklet from, 77
Wiring: and door retrofits, 30
Wood Handbook (USDAFPL): mentioned, 38
Workbenches: for door work, 54, 55
Wrecking bars: for trim removal, 44

The articles in this book originally appeared in *Fine Homebuilding* magazine. The date of first publication, issue number and page numbers for each article are given at right.

8 **Choosing a Front Door**
August 1994 (90:40-45)

14 **Connecticut River Valley Entrance**
October 1982 (11:36-41)

20 **A Greek Revival Addition**
June 1989 (54:68-71)

24 **Formal Entryway**
June 1981 (3:14-18)

29 **French-Door Retrofit**
June 1991 (68:42-46)

34 **Trimming the Front Door**
February 1993 (79:46-49)

38 **Making an Insulated Door**
August 1982 (10:46-47)

40 **Batten Doors**
February 1982 (7:32-33)

42 **Replacing an Oak Sill**
February 1984 (19:34-36)

45 **A Breath of Fresh Air**
December 1986 (36:39-41)

48 **Building Wooden Screen Doors**
December 1989 (57:72-75)

52 **Production-Line Jamb Setting and Door Hanging**
April 1989 (53:38-42)

57 **Hanging an Exterior Door**
April 1982 (8:34-36)

60 **Ordering and Installing Prehung Doors**
April 1992 (74:62-65)

64 **Installing Prehung Doors**
June 1995 (96:46-51)

70 **Hanging Interior Doors**
April 1985 (26:26-32)

77 **Casing a Door**
December 1985 (30:55-57)

80 **Building Interior Doors**
February 1992 (72:50-53)

84 **Pueblo Modern**
April 1991 (67:56-58)

87 **Pocket Doors**
June 1989 (54:63-67)

92 **Curved Doors**
December 1982 (12:61-63)

95 **An Elegant Site-Built Door**
August 1993 (83:57-61)

100 **Installing Mortise Locksets**
April 1993 (81:60-63)

104 **Installing Locksets**
February 1993 (79:40-45)

110 **Homemade Hardware**
October 1990 (63:66-68)

113 **Door Hardware**
August 1988 (48:61-65)

118 **More Than a Door**
August 1995 (97:96-97)

120 **A Welcome Home**
October 1989 (56:88-89)

122 **Carved Doors**
April 1988 (46:84-85)

124 **Corner Copia**
October 1988 (49:90-91)